The Pro Tools 2023 Post-Audio Cookbook

A holistic approach to post audio workflows like music production, motion picture, and spoken word

Emiliano Paternostro

BIRMINGHAM—MUMBAI

The Pro Tools 2023 Post-Audio Cookbook

Group Product Manager: Rohit Rajkumar

Publishing Product Manager: Chayan Majumdar

Senior Content Development Editor: Feza Shaikh

Technical Editor: Simran Udasi

Copy Editor: Safis Editing

Project Coordinator: Sonam Pandey

Proofreader: Safis Editing

Indexer: Manju Arasan

Production Designer: Prashant Ghare

Marketing Coordinators: Anamika Singh, Namita Velgekar, and Nivedita Pandey

First published: August 2023

Production reference: 1050723

Packt Publishing Ltd
Grosvenor House
11 St Paul's Square
Birmingham
B3 1R

ISBN 978-1-80324-843-1

www.packtpub.com

This book is dedicated to David Chellew. His steadfast determination inspired many people to follow their hearts and his boundless kindness touched them.

– Emiliano Paternostro

Contributors

About the author

Emiliano Paternostro is an Avid-certified Pro Tools Instructor who has worked with audio for more than 25 years. He has worked on countless projects in various capacities and continues to provide post-audio work for podcasts and long-form content. He continues to expand his skill set and is currently researching machine learning tools for cleaning audio signals. You can find a list of his work at www.proximitysound.com.

This book would not have been possible without the unwavering support of my wife, Shannon, to who I owe my unending gratitude.

About the reviewers

Yanni Caldas is a Canadian audio specialist based in the Toronto area. With a focus on immersive storytelling through sound design and implementation, Yanni works independently under his company, AmnesiaSound. His past projects include sound design for AA and indie game studios, composition for national television and broadcast, mixing, production, and recording for major and indie musical acts, and audio restoration for forensics. As an active field recordist, Yanni participates in crowdsource libraries that often contribute to various charities. As he continues to nurture his abilities, what remains is Yanni's fundamental passion for sound: *"Just like photos capture moments, audio recordings capture feelings."*

To learn more about Yanni's work, visit his website www.amnesiasound.ca

Marisa Ewing (she/they) is a dialog editor, sound designer, and mixing/mastering engineer. She edits audio for podcasts, films, and video games, and has previous experience in both live music and corporate audio-visual work. She has been using Pro Tools for over a decade.

In 2020, Marisa founded Hemlock Creek Productions, a production company focused on audio editing needs for remote productions. In 2023, Hemlock Creek Productions released its first original show, the horror audio drama *Liars & Leeches*, where Marisa worked as an audio engineer, as well as being the show's creator and director.

Some of Marisa's past projects include the video game *The Last of Us Part 1*, as well as the podcasts *Maxine Miles* and *Dark Dice*.

To learn more about Marisa's work, visit her website www.hemlockcreekprod.com

Table of Contents

2

Importing and Organizing Audio   29

3

Faster Editing Techniques   55

4

The Mechanics of Mixing   81

5

Shaping Sounds with Plugins and Effects   109

6

Finishing a Project and Creating Deliverables   145

7

Considerations for Music Production   179

8

Post Production for Motion Pictures   199

9

Preface

Since its release in 1991, Pro Tools has been ubiquitous in professional audio environments by providing a powerful software interface with high-quality audio hardware. Even after removing the hardware requirement in 2010, Avid's digital audio workstation and its accompanying hardware options are seen throughout the professional audio world in music studios and mix theatres for motion pictures. While there are a myriad of audio tools out there, Pro Tools still commands a strong presence in the world of audio, and the ability to use it well can help bolster your path into the industry.

If you ask 100 different Pro Tools users what the best way to perform a specific task is, you'll get 100 different answers. Pro Tools is powerful in its flexibility but creates a bit of a dilemma when trying to figure out what workflows fit best for certain situations. This book is not *the* way to use Pro Tools, but *my* way.

Instead of trying to show you all the tools and how they operate with their nuances, I've taken a typical project workflow from start to finish and written out how I do it, focusing on the core competencies. In this book, you'll find the different ways I've learned to get the most out of Pro Tools to complete projects in a fast and efficient manner.

After covering the major aspects of a project workflow, you'll get some tips and tricks for working with specific mediums. The goals and processes for audio differ depending on whether you're working on motion pictures, music, or podcasts and spoken word, so the last chapters of this book focus on ways to work within those modes.

Pro Tools can be daunting to begin with, and even after decades of working with it, I still learn new things about it every day. I hope you'll find new ways to look at your projects and faster techniques through the recipes outlined in this book and have a little fun along the way.

Who this book is for

This book is intended for audio professionals and students who use Pro Tools as their main digital audio workstation.

Examples of roles in the audio world that could benefit from the recipes in this cookbook are as follows:

- Sound editors for linear media, such as motion pictures, podcasts, and audiobooks
- Sound mixers and re-recordist mixers
- Music engineers and producers

What this book covers

Chapter 1, Planning and Preparing Sessions, shows what goes into a Pro Tools project before any audio is brought into it, including track layouts and organization.

Chapter 2, Importing and Organizing Audio, takes you through ingesting media into a Pro Tools session and organizing audio once imported.

Chapter 3, Faster Editing Techniques, shows you how to quickly traverse audio clips and use the keyboard to edit projects with ease.

Chapter 4, The Mechanics of Mixing, takes you through the technical side of the mixing phase of a project from using automation to routing audio via aux tracks.

Chapter 5, Shaping Sounds with Plugins and Effects, covers the creative side of mixing and the tools used to manipulate how something sounds.

Chapter 6, Finishing a Project and Creating Deliverables, examines how a project gets exported when it's done and what to consider for project delivery.

Chapter 7, Considerations for Music Production, takes you through the nuances of music production and mixing.

Chapter 8, Post Production for Motion Pictures, takes you through how Pro Tools is used for motion picture projects with an emphasis on different dialogue-editing techniques.

Chapter 9, Spoken Word and Podcasts, offers methods for quickly producing spoken word projects, such as podcasts and audiobooks.

To get the most out of this book

While this book will provide you with step-by-step instructions for completing the recipes, having a basic understanding of digital audio workstations and how to operate Pro Tools will be beneficial. This book will not guide you through installing and setting up Pro Tools, for example.

Software/hardware covered in the book	OS requirements
Pro Tools 2023.3	One of the following:
	Windows 10
	Windows 11
	macOS Catalina (10.15.7)
	macOS Big Sur (11.7.4)
	macOS Monterey (12.6.3)
	macOS Ventura (13.2.1)

Please note that certain plugins mentioned in the book are not accessible in the trial version of Pro Tools Ultimate. Kindly make use of the specified version mentioned in the Technical Requirements section of the chapter.

Download the example code files

You can download the example sessions and audio files for this book from GitHub at `https://github.com/PacktPublishing/The-Pro-Tools-2023-Post-Audio-Cookbook`. In case there's an update to these files, it will be updated on the existing GitHub repository.

We also have other code bundles from our rich catalog of books and videos available at `https://github.com/PacktPublishing/`. Check them out!

Download the color images

We also provide a PDF file that has color images of the screenshots/diagrams used in this book. You can download it here: `https://packt.link/Zwua2`.

Conventions used

There are a number of text conventions used throughout this book.

`Code in text`: Indicates code words in text, database table names, folder names, filenames, file extensions, pathnames, dummy URLs, user input, and Twitter handles. Here is an example: "Name the rest of the submixes `Effects Submix`, `Foley Submix`, `Music Submix`, and `Ambience Submix`."

When we wish to draw your attention to a particular part of a code block, the relevant lines or items are set in bold:

Bold: Indicates a new term, an important word, or words that you see onscreen. For example, words in menus or dialog boxes appear in the text like this. Here is an example: "In the **Menu** bar, select **Setup | Session**."

> **Tips or important notes**
> Appear like this.

Sections

In this book, you will find several headings that appear frequently (*Getting ready, How to do it..., How it works..., There's more...,* and *See also*).

To give clear instructions on how to complete a recipe, use these sections as follows:

Getting ready

This section tells you what to expect in the recipe and describes how to set up any software or any preliminary settings required for the recipe.

How to do it...

This section contains the steps required to follow the recipe.

How it works...

This section usually consists of a detailed explanation of what happened in the previous section.

There's more...

This section consists of additional information about the recipe in order to make you more knowledgeable about the recipe.

See also

This section provides helpful links to other useful information for the recipe.

Get in touch

Feedback from our readers is always welcome.

General feedback: If you have questions about any aspect of this book, mention the book title in the subject of your message and email us at customercare@packtpub.com.

Errata: Although we have taken every care to ensure the accuracy of our content, mistakes do happen. If you have found a mistake in this book, we would be grateful if you would report this to us. Please visit www.packtpub.com/support/errata, selecting your book, clicking on the Errata Submission Form link, and entering the details.

Piracy: If you come across any illegal copies of our works in any form on the Internet, we would be grateful if you would provide us with the location address or website name. Please contact us at copyright@packtpub.com with a link to the material.

If you are interested in becoming an author: If there is a topic that you have expertise in and you are interested in either writing or contributing to a book, please visit authors.packtpub.com.

Reviews

Please leave a review. Once you have read and used this book, why not leave a review on the site that you purchased it from? Potential readers can then see and use your unbiased opinion to make purchase decisions, we at Packt can understand what you think about our products, and our authors can see your feedback on their book. Thank you!

For more information about Packt, please visit packtpub.com.

Share Your Thoughts

Once you've read, we'd love to hear your thoughts! Scan the QR code below to go straight to the Amazon review page for this book and share your feedback.

https://packt.link/r/1803248432

Your review is important to us and the tech community and will help us make sure we're delivering excellent quality content.

Download a free PDF copy of this book

Thanks for purchasing this book!

Do you like to read on the go but are unable to carry your print books everywhere? Is your eBook purchase not compatible with the device of your choice?

Don't worry, now with every Packt book you get a DRM-free PDF version of that book at no cost.

Read anywhere, any place, on any device. Search, copy, and paste code from your favorite technical books directly into your application.

The perks don't stop there, you can get exclusive access to discounts, newsletters, and great free content in your inbox daily

Follow these simple steps to get the benefits:

1. Scan the QR code or visit the link below

https://packt.link/free-ebook/978-1-80324-843-1

2. Submit your proof of purchase
3. That's it! We'll send your free PDF and other benefits to your email directly

1

Planning and Preparing Sessions

Before opening Pro Tools or any **Digital Audio Workstation** (**DAW**), it's important to set up your project for success. A well-planned, prepared, and laid-out session will save you hassle and frustration down the line when you're trying to locate and manipulate sounds within the mix. Knowing the ins and outs of track routing and management will mean you can focus more of your time on the creative elements of your project, and you'll also be able to confidently share your sessions with collaborators, knowing they'll understand the session structure. Finally, a well-structured session will ensure that, should you revisit a project, you'll be easily able to pick up where you left off. You may find yourself having to revisit a session years after the fact or refer to it for another project, and you'll thank yourself for your diligence.

In this chapter, we'll go over planning your audio project, setting up a session to the correct specifications, and then utilizing the track and routing tools available. We'll also examine some of the organizational tools, such as track groups and markers.

We'll cover the following recipes:

- Which version of Pro Tools is right for you?
- Planning the project on "paper" first
- Setting up the session
- Creating tracks
- Setting up aux tracks for audio routing
- Organizing with Folder Tracks
- Grouping tracks for editing, mixing and viewing
- Using memory locations within a project

Technical requirements

This chapter requires at least Pro Tools Intro installed. Specific recipes will note when higher versions are required.

The example sessions and audio files for each recipe can be found at `https://github.com/PacktPublishing/The-Pro-Tools-2023-Post-Audio-Cookbook/`.

Which version of Pro Tools is right for you?

Pro Tools has seen many changes over its over-30-yearear history. There are currently four versions of Pro Tools available, and you may not know which one is right for you and your workflow. Avid (the maker of Pro Tools) does provide detailed charts that can help you figure out which route to take, but these can be difficult to parse, especially if you haven't used Pro Tools before, or even if you have stuck with an older version for a while. Since moving to a subscription model, it is also difficult to find older versions of Pro Tools and their licenses available, so we're going to break down the four versions of Pro Tools available at the time of writing and determine which one would work best for you. When looking at the costs listed here, keep in mind that Avid also offers discounts for yearly subscriptions, which can be sizable, depending on the version.

Pro Tools Intro

The newest addition to the Pro Tools family, Pro Tools Intro is Avid's free version of Pro Tools that offers a very small slice of what can be done with the software. While the limits can be very offputting for experienced professionals (including no video support), the number of plugins included is quite versatile and Pro Tools Intro could be perfectly suitable for many smaller audio-only projects. With Intro, you get the following:

- Eight tracks each of audio, instruments, and MIDI
- 36 plugins
- Four aux tracks
- One master track
- Up to four tracks of simultaneous recording
- Stereo output only
- Forum support only

If you are just beginning to use Pro Tools or DAWs in general, this version could be a good place to start. Limitations can be beneficial as being exposed to all possible tools can become overwhelming. There's no financial commitment, so there's the incentive to continue using it and not be burdened by cost. Pro Tools Intro can also be a great solution for voice actors recording at home, or podcast producers wanting more fine-tuned editorial features while not being encumbered by features you may not need.

Pro Tools Artist

At the lowest paid tier is Pro Tools Artist. For $9.99 per month, you get a more functional DAW, but still no video support. It includes the following:

- 32 audio and instrument tracks
- 64 MIDI tracks
- 113 plugins
- 32 aux tracks
- 1 master track
- Up to 16 tracks of simultaneous recording
- Stereo output only
- Standard support

Artist provides plenty of power for simple music production and spoken word audio. The plugins included are also quite versatile and getting standard support for issues beats having to rely on forum support. This would be a good choice for those wanting to expand on their audio production skills but who don't need more advanced tools.

Pro Tools Studio

For $29.99 per month, you can get the full toolset experience from Avid with Pro Tools Studio. This version of Pro Tools is extremely capable for almost all audio production scenarios and offers these things:

- 512 audio and instrument tracks
- 1,024 MIDI tracks
- 125 plugins
- 128 aux tracks
- 64 master tracks
- Up to 64 tracks of simultaneous recording

- 128 VCA tracks

- One video track

- Stereo, surround, Dolby Atmos, and Ambisonics output

- Standard support

If you're working on an audio post for a motion picture, or want to work with multichannel audio beyond stereo, then you need to have at least Pro Tools Studio.

Pro Tools Ultimate

The highest tier from Avid is Pro Tools Ultimate, coming in at $99.99 per month. If you are working with large-scale productions or need advanced audio post tools, then you'll want to consider this version. Pro Tools Ultimate is also required for many of Avid's hardware acceleration products, such as their HDX and HD Native systems. You can, however, run Pro Tools Ultimate without extra hardware support. Ultimate provides the following:

- 2,048 audio tracks

- 512 instrument tracks

- 1,024 MIDI tracks

- 125 plugins

- 1,024 aux tracks

- 512 master tracks

- Up to 256 tracks of simultaneous recording (depending on hardware)

- 128 VCA tracks

- 64 video tracks

- Stereo, surround, Dolby Atmos, and Ambisonics output

- ExpertPlus support (phone support and higher priority/case response time)

If you are working professionally in audio, then these comparisons should help you decide which version will work best for your needs.

There's more...

While this book is aimed at Pro Tools users, it's important to acknowledge that in the end, the DAW you use is just a tool. There are many different DAWs available that focus on different areas of audio production. Almost every DAW comes with a trial period, so you can take them for a test run and see what features and interface work best for you. I also recommend speaking with other audio production professionals and users and getting their opinion on what tools work best for which scenarios and what might work best for you.

Here are some examples of other DAWs you could consider:

- Reaper

- Nuendo

- Digital Performer

- Studio One

- Ableton

- Logic

- Ardour

See also

Avid also offers a detailed comparison of all the features on its website: `https://www.avid.com/pro-tools/comparison-extended`.

Planning the project on "paper" first

The biggest mistake I see from early users of Pro Tools or any DAW, in general, is diving right into a project without a solid plan. After becoming comfortable with the inner workings of a DAW, you may be able to jump right into a project with a basic idea of what's going to be needed, but there are many advantages to determining the scope and requirements of a session before even opening your DAW. Let's look at some examples of what works for different types of projects.

Scripts

If you are working on story-based content, then a script is a great place to start. This might seem like extra work for some non-fiction projects – why not simply bring in all the audio and start editing it in the DAW? The reality is that audio clips and waveforms without any pre-planned guidance are much more challenging to cultivate into a cohesive narrative from within Pro Tools and almost every other DAW. Some of the most powerful audio software out there treats the spoken word as text and encourages editing it that way (check out *Hindenburg*, *Descript*, and *iZotope RX*). Here's an example script from the *Twenty Thousand Hertz* podcast:

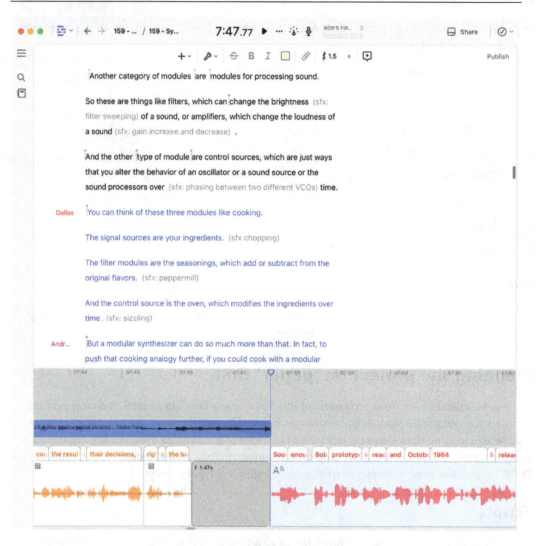

Figure 1.1: A Twenty Thousand Hertz podcast Descript project

For this podcast, the different elements are formatted to make it easier to parse. The name of the person speaking is shown on the side, and audio cues for sound effects and music are in brackets and colored gray. The guest audio was not scripted; this was transcribed from an interview first and then decisions were made for which parts would be used in the show. This was all done on "paper" before any audio was edited. The narration was written for the host and all the sounds were also inserted into the script so that the entire team could be on the same page. Whether you are working with a large team or just by yourself, having everything written out first makes it much easier to plan out the project and determine what will be needed.

Spreadsheets

One step further from a script is a spreadsheet. Sometimes called a cue sheet, this takes the information from the script and breaks it down from moment to moment to lay out all the audio materials needed for the project. This is always the first step in any of the motion picture projects for which I have done sound design work. I have a "spotting session" with the director and go moment to moment throughout the piece and determine what should happen at every "spot." Here is a sheet for a recent project I worked on, Christopher Walsh's *Orchid*:

Timecode	Type	Description	Notes
01:00:00:00	Ambience	Opening Title	Sounds of the city - traffic, far off sirens, horns, dog barks. Cheap apartment, radiator, ch
01:00:05:16	Ambience	Apartment	Sounds of the city - traffic, far off sirens, horns, dog barks. Cheap apartment, radiator, ch
01:00:39:16	Ambience	Outside Bar	Music (provided by composer), light traffic
01:01:04:10	Ambience	Alley	Light traffic and music gets muffled
01:01:14:22	Effect	Alley	Gun shot
01:01:25:21	Effect	Alley	Soul escapes - Passing through a portal, throbbing sound?
01:01:45:04	Ambience	The Lab	Electronic Equipment, Sparks, Bubbling, Humming
01:02:36:17	Effect	The Lab	The Machine powers ups
01:03:25:00	Ambience	Apartment	Sounds of the city - traffic, far off sirens, horns, dog barks. Cheap apartment, radiator, ch
01:03:35:17	Music	Dream	Only Music
01:04:16:18	Ambience	Apartment - Night	Middle of the night, city sounds, but quieter and less frequent
01:04:26:10	Ambience	Apartment - Day	Sounds of the city - traffic, far off sirens, horns, dog barks. Cheap apartment, radiator, ch
01:04:42:00	Ambience	Playground	Kids playing
01:05:12:05	Effect	Playground	Recess Bell
01:07:50:06	Ambience	Apartment	Sounds of the city - traffic, far off sirens, horns, dog barks. Cheap apartment, radiator, ch

Figure 1.2: Orchid's sound design log

Having detailed breakdowns of the needs of a project can help immensely. Knowing what sound effects you will need to either source from a library or capture/create means you can better plan your sessions. This also helps you gain a scope of the project when trying to determine what costs might be in terms of labor and time that needs to be put in.

Track layouts

Up until this point, I've detailed techniques that are typically used in motion picture post audio and spoken word or podcasts, but what if your project is more musical? A "script" of what the listener's journey should be could still be very useful. Before the advent of DAWs and automation, songs and mixes would have a script of sorts that would detail what the mixer(s) would need to do when it came time to bounce and print the final product. While there is still great benefit to this practice, I find it especially useful to draw out a track routing layout to help me visualize what the project will look like. Here's an example from a song I mixed for the artist *SMBRZ*:

VOX - MAIN			[SEND TO VERB A]
	Vocal - Main	*slight saturation/OTT*	
	Vocal - Main Double		
	Vocal - Harmony		
	Vocal - Harmony Double		
	Vocal - Harmony 2		
VOX - BACKUP			[SEND TO VERB A]
	Vocal - Background 2	*phaser/flange*	
	Vocal - Background 2		
	Vocal - Background 3		
STRINGS			
	Guitar 1	*hard left*	
	Guitar 2	*hard right*	
	Guitar 3	*center*	
	Guitar 4	*center*	[SEND TO DELAY]
	Guitar 4 Harmony	*center*	
BASS			
	Bass - Electric	*punchy, slap*	
	Bass - Synth	*low end emphasis*	
DRUMS			[OUT TO COMP]
	Overhead Left		
	Overhead Right		
	Floor Tom		
	Toms		
	Snare - Top		
	Snare - Bottom		
	Kick	*very tight, dry*	

Figure 1.3: Sample track layout chart

The design of a track layout chart or map is dependent on the project and the way you like to visualize things. Color coding is useful for grouping similar tracks, and it can also be useful to add routing notes. The overall goal is to get a sense of what the project will look like as a whole and where things should go. Many times, when I received tracked songs from independent artists, there would be no rhyme or reason to their track layout, and they would simply add tracks as more instruments were needed. This made it very difficult to parse what was happening and communicate with the artist about their needs.

There's more...

The most important takeaway from this concept is that planning and writing everything down before creating your sessions will make things easier in the long run. You don't need to commit to your plan, the project will often change, and new requirements will need to be considered, but a visualization of where things should go will help you keep things on track. Finally, making changes on paper is much easier and cheaper than trying to implement them during a project.

Setting up the session

Before you can begin doing any work in Pro Tools, you need to set up a session. Some aspects of a session can be changed after the fact, and some cannot. Knowing the different session formats is important, as well as understanding how to set them up correctly. For this walkthrough, we'll create a generic session and discuss the different options and their impact/effects.

Getting ready

If you are working with specific audio files, look at what sample rate they are in. On macOS, you can find this out by selecting the file in **Finder** and going to **File | Get Info**. On Windows, you can install a third-party tool such as **MediaInfo**. This will allow you to right-click on the file in File Explorer and select it as an app to open the file with. The app will show you the sample rate, along with other file metadata.

How to do it...

For this recipe, we will be creating a sample project with a sample rate of 44.1 KHz and 24-bit Bit Depth from Pro Tool's **Dashboard** window.

Let's do that by following these steps:

1. Open Pro Tools. If you are not greeted with the **Dashboard** window, click **File | Create New...**

2. On the tab to the left, select **Create**.

3. Under **Name**, type `Example Session`.

4. For the session type, select **Local Storage (Session)**.

5. Deselect **Create From Template**.

6. For **File Type**, select **BWF (.WAV)**.

7. For **Sample Rate**, select **44.1 kHz**.

8. For **Bit Depth**, select **24-bit**.

9. For **I/O Settings**, select **Last Used**.

10. Deselect **Interleaved**.

11. Select **Prompt for location**.

12. Click the **Create** button:

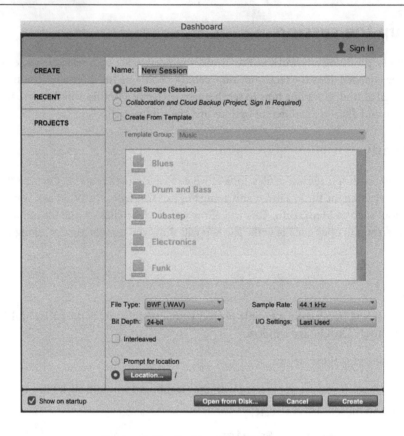

Figure 1.4: The Dashboard's CREATE pane

How it works...

Let's go over all the different project settings and their effects.

Session Type

Avid offers cloud storage and collaborative tools for the higher tiers of Pro Tools, which it refers to as Projects. A Local Storage Session is not able to access those options and will store all the files locally on your computer.

Create from Template

Once you are comfortable using Pro Tools, you can save your sessions as templates to be selected from this menu or opened within the OS. If you want a blank session, make sure this is unchecked.

File Type

Most audio software these days uses **Broadcast Wave Format** (**BWF**), often simply called WAV files; these have the WAV file extension. AIFF is Apple's lossless audio format introduced in the late 90s. For a time, using AIFF was more stable than using Mac software, but currently, BWF is more compatible with most professional post-production software.

Sample Rate

The options available depend on your hardware and audio interface. Higher sample rates are a good choice for sounds that will be slowed down later and to avoid harmonic distortion with specific tools, but other issues can be introduced when converting to the target sample rate at the end of the project. Many audio plugins also oversample to mitigate these issues. Traditionally, **44.1 kHz** is used for CD audio, and **48 kHz** is used for motion picture. Checking your source material is the best way to avoid issues. If your project matches the source audio, you won't need to convert the files before working with them.

Bit Depth

The higher the bit depth, the more dynamic range your project will have. The calculation is approximately 6 dB per bit, so a **16-bit** file has **96 dB** of depth, whereas a **24-bit** file has **144 dB**. This might not seem important, but to be able to get strong impactful sound, you need headroom, and limiting the amount of depth a session has to work with will make things difficult to mix effectively. It is also possible to mix at **32-bit (float)**, which gives virtually limitless headroom, with the drawbacks of larger file size and potential quantizing errors when exporting later.

I/O Settings

Like session templates, **I/O Settings** can be used to pre-set certain routing in a session.

Interleaved

Interleaved audio is what most users are accustomed to when seeing audio files on their computers. Multiple channels of audio (such as left and right for stereo files) are combined into a single file. Interleaved audio is easier to share with collaborators and more convenient to pull from the project's session folders if needed, but the alternative, which is separate files for each channel, is often needed when providing files for professional delivery. This checkbox will rarely impact your usage of Pro Tools, and ultimately, you can always export files to a different format later.

Prompt for Location

This will force you to pick a location on your system to save the file. If you try to save your session to a drive that is not fully supported (such as Google Drive), you will get a warning that it has to be saved to a **recordable** (**R**) drive. You will need to pick another place to save it. You can also select the location it will be saved to in advance with the option below it, but be careful – forgetting this option will often lead to frustration if you forget where your session has been saved.

There's more...

You can change some options for a session after the fact. With a session open, go to the menu bar and select **Setup | Session** (*Command + 2* on macOS, *Ctrl + Numpad 2* on Windows); you will see that almost all options can be changed except for **Sample Rate**. This is why it's important to set this correctly when creating a session.

Creating tracks

Now that you have set up a session, it's time to populate it with some tracks. Pro Tools offers up to nine different tracks, each with functions and purpose. We'll create some simple tracks with two different methods and explain their purpose in brief.

Getting ready

For this recipe, you will need a blank session. You may use the Example session you created in the previous recipe. In your Pro Tools session, make sure that **Track Color** is enabled in the **Edit** window. You can do this by going to the menu bar, selecting **View | Edit Window Views**, and making sure it is checked or using the **Edit Window View** selector dropdown directly above the track headers on the left-hand side of the **Edit** window (see *Figure 1.5*):

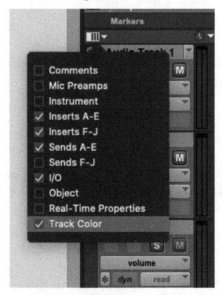

Figure 1.5: The Edit Window View selector dropdown

How to do it...

The **New Track** window can almost set up an entire session in one shot by allowing you to create a variety of tracks and customize their order. Track names are sequentially incremented, so if you have something such as [Guitar 1, Guitar 2, Guitar 3], you can simply create three tracks named `Guitar` and it will add the numbers for you.

For this recipe, we'll create 12 tracks in our session. These will be different types, both in channel count and function. We'll also color-code the tracks for better organization. Here are the steps to do so:

1. In the menu, select **Track** | **New...** (*Command* + *Shift* + *N* on macOS, *Ctrl* + *Shift* + *N* on Windows).

2. For **Create**, type 4.

3. For the channel count, select **Stereo**.

4. For the track type, select **Audio Track**.

5. Leave the third dropdown set to **Samples**.

6. For **Name**, type `Stereo Audio Track`.

7. Click the + button to the right.

8. For the next row, follow *steps 1-7*, but change the channel count to **Mono** and the name to `Mono Audio Track`.

9. Create three more rows. For these rows, create **1 Stereo Aux** track, **1 MIDI** track, **1 Stereo Instrument**, and **1 Master Fader** track and name them accordingly.

10. Click the **Create** button:

Figure 1.6: The New Tracks window

11. Go to **Window** | **Color Palette** or double-click the colored space on the left side of the **Track** header.

12. In the **Color Palette** window, at the top left, click the dropdown and select **Tracks** (see *Figure 1.7*):

Figure 1.7: The Color Palette window with Tracks selected

13. Click on the name of the first track (**StereoAudTrck1**), then hold *Command* (*Ctrl* on Windows), and click on the three other stereo audio tracks.

14. Choose a red color from the **Color Palette** window.

15. Click on the name of the first mono audio track, hold *Shift*, and click on the last mono track.

16. Choose a purple color in the **Color Palette** window.

How it works...

Five main types of tracks are used in Pro Tools. Most tracks can be measured in either samples or ticks. Samples are finite. They directly correspond to a measurement of time and the sample rate of the session. If you have a sample rate of 48 kHz, then you know that each sample is 1/48,000 of a second. Ticks are fluid. They are not directly related to time, but rather the space between beats in a bar of music. Pro Tools provides 960 ticks per beat. Depending on the tempo of the music, a tick can be very short or very long. For example, if a song has 60 **beats per minute** (**BPM**), then each beat is 1 second long and has 960 ticks between them. So, one tick is 1/960th of a second. If a song is 120 BPM, then there are two beats per second, doubling the number of ticks per second and increasing the speed of a tick to 1/1,920th of a second. The important thing to note is that increasing or decreasing the tempo of a session will not impact tracks measured in samples, but will impact tracks measured in ticks (they will speed up or slow down).

Audio tracks

Audio tracks hold WAV audio clips. Any files that consist of sampled audio data (as opposed to MIDI instructions) can go here. Tracks can be mono, stereo, or any number of channels that your version of Pro Tools can support. These are typically measured in samples.

Aux (auxiliary) tracks

Aux tracks cannot hold any audio data but instead are used as a way of routing audio for effects or for summing/mixing down signals from other tracks. Other tracks can have their output sent directly to an Aux or can be "split" and sent separately through a send. They can also be mono, stereo, or multi-channel, like audio tracks. These are also typically measured in samples.

MIDI tracks

MIDI tracks hold MIDI note data. They can record MIDI information and can be used to edit MIDI sequences, but they cannot be used to generate sounds. MIDI tracks need their data signal sent to an Aux track with an instrument plugin placed on an insert, or to external MIDI hardware. They cannot have channels such as mono or stereo. These are typically measured in ticks.

Instrument tracks

Instrument tracks are a combination of a MIDI track and an Aux track. A software instrument plugin still needs to be inserted into this track to generate information, but MIDI note data can be directly written to these tracks. These are usually measured in ticks.

Master tracks

Master tracks are the last step before outputting to your hardware. Unlike other tracks, plugins and effects placed on the inserts here are *post-fader*. I generally avoid using master tracks, but they can be useful for setting master levels while not affecting headroom, or for getting proper level/analysis on signals.

You can tell how many channels a track has by looking at its VU meters, which can be found on the right-hand side of the **Track** header. The mono tracks we created have only one meter, while the stereo tracks have two. Multi-channel tracks such as surround will also show more VU meters – six for 5.1 tracks, for instance.

Color-coding the tracks makes it much easier to see the types of tracks you have in your session, and your color-coding scheme can follow any convention you see fit for your project. Selecting multiple individual tracks with *Command* + click (*Ctrl* + click for Windows) and ranges of tracks with *Shift* + click is a great way to apply changes to multiple tracks at the same time. You can also *Option/Alt* + click on a track to deselect them all.

There's more...

At any time, you can click the arrow at the bottom left of the **Edit** window to bring the tracklist in and out and see all your tracks from a bird's-eye view (see *Figure 1.8*). You can also click and drag the name of tracks from the tracklist or from within the **Edit** window to rearrange tracks. If you accidentally made too many tracks, you can right-click and select **Delete** to remove them. If there are audio clips on the track, Pro Tools will warn you and ask you to confirm before the tracks are deleted as you cannot undo this command:

Figure 1.8: The Track sidebar's Show/Hide button

There are some shortcuts for adding tracks to your edit window as well. In any empty area below a track, double-click to add another track of the same configuration as your most recently added track. If you want a different type of track, try holding the following modifier keys on your keyboard while double-clicking on an empty area:

- *Command*: Add an audio track (*Ctrl* on Windows)
- *Option*: Add an instrument track (*Alt* on Windows)
- *Control*: Add an Aux track (the *Windows* key on Windows)
- *Shift*: Add a master track

Setting up Aux tracks for audio routing

Auxiliary tracks, known as Aux tracks in Pro Tools, are a vital part of signal flows. Aux tracks don't contain any audio data and clips cannot be added to a track, but they can do pretty much anything else an audio track can do. Plugins can be added as inserts, volume and pan information can be adjusted, and Aux tracks can be routed and sent to other tracks. Besides not holding audio data, Aux tracks always have input monitoring enabled, so sound always flows to them. You can use Aux tracks as a way of summing track signals together for better control, or as a place to send signals to when you want to redirect the sound to different tracks.

Getting ready

This recipe requires a Pro Tools session with at least four audio tracks. It helps to have some audio clips in there to see the effect. In your Pro Tools session, make sure that you have both the **I/O** and **Sends** columns showing in the **Edit** window. You can do this by going to the menu bar and selecting **View | Edit Window Views** and making sure the appropriate items are checked (see *Figure 1.9*), or by using the **Edit Window View** selector dropdown directly above the track headers on the left-hand side of the **Edit** window:

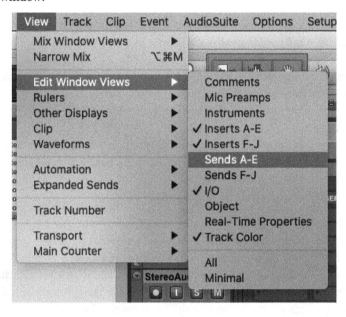

Figure 1.9: Enabling Sends and I/O from the Edit Window Views menu

How to do it...

For this recipe, we'll create two Aux tracks in a session and route audio to them at the same time:

1. Go to the **Edit** window, select a track, and click on the output dropdown in the **I/O** column.

2. Click on **New Track...** at the bottom.

3. Leave the default parameters at **Stereo Aux Input in Samples** named Aux 1 but uncheck **Create next to current track**. Then, click the **Create** button.

4. Click on another track and set its output to the same aux track by clicking the output dropdown in the **I/O** column and selecting **Track | Aux 1**.

5. In another track, go to the **Sends A-E** column and click on the slot with the letter **a**.

6. Select **New Track...** and use the same parameters as above to make another Aux track named Aux 2.

7. Click the button for **Send A** labeled **Aux 2** to have a fader appear.

8. Press *Option/Alt* + click on the fader to bring its level to 0.

9. In the **Edit** window, hold *Option/Alt* and click and drag the button for **Send A** labeled **Aux 2** to the **Send** slot labeled **a** on the next track.

How it works...

We've created and used Aux tracks in different ways, so let's break them down. The first method was to create an Aux track by simply sending the track's output to a **New Track**. We used the generic name Aux 1 for simplicity, but this could represent a mix-down for any number of audio tracks. Perhaps you want to sum up all the signals from a drum kit so that you can increase or decrease its overall loudness with one fader/channel. Or perhaps you want to mix down all the dialog tracks in a podcast so you can make sure they are monitored at a different isolated level. Using an Aux track is one way to accomplish that. The key is to set the output of an audio track to that Aux track. This means 100% of the signal will go there and nowhere else. It is possible to have multiple Aux tracks use the same bus/input, but we'll avoid doing that for now. Once an Aux track has been created, you can simply set other tracks to it using the **Output** dropdown.

The second method of using an Aux track was by using a Send. This is drastically different as the signal from the audio track is still being sent to the main output (your speakers) but is also being split and sent to the Aux simultaneously. Once a Send is set, its volume is set to infinite (essentially muted) by default, so you need to set its output using the fader that appears when you click on it. If you are happy with a Send level's setting, then you can *Option/Alt* + click and drag it to duplicate it to another track. Using Sends to split tracks can be useful for auditioning different effects on different Aux tracks, blending and mixing multiple effects, or creating different sets of prints/deliverables:

Figure 1.10: Sending an audio track to multiple Aux tracks to audition different plugins

There's more...

Aux tracks follow the same conventions as other tracks when using Solo modes. That is to say, if you Solo a track that has its output routed to an Aux track, the aux track will be muted, and you won't hear anything. You can try to solo both the audio track and the aux track, but the better method is to **Solo Safe** the Aux track. You can do this by holding *Command* (*Ctrl* on Windows) and clicking on the **Solo** button in the track (the **S** button below the track's name). You will notice it becomes grayed out. This track will now play, regardless of another track being in Solo mode. By default, Sends are also post-fader. The signal they output will be affected by the track's actual volume after the fader is adjusted. If you wish to have the Send's signal not affected by the main fader, click the **PRE** button on the Send's fader.

Organizing with folder tracks

Folder tracks are a relatively new concept to Pro Tools, having been introduced in version 2020.3, and as such are not always utilized as heavily as other organization tools. I love folder tracks and use them heavily in my projects, especially in music. Folder tracks can be used purely for organization and have no routing capabilities or be used as routing folders that allow tracks to have their signals routed to them. You can also nest folders within folders if you want:

Figure 1.11: Nested folder tracks

Getting ready

You will need a Pro Tools session with a mix of mono, stereo, and other tracks added to it for this recipe. In your Pro Tools session, make sure that the **I/O** columns are shown in the **Edit** window. You can do this by going to the menu bar and selecting **View | Edit Window Views** and making sure it is checked or using the **Edit Window View** selector dropdown directly above the track headers on the left-hand side of the **Edit** window.

How to do it...

In this recipe, we'll create two folder tracks and organize other tracks into them. We'll also route the audio to those tracks appropriately. Follow these steps:

1. Go to **Track | New...**.
2. Configure the settings to create **2 Stereo Routing Folders** and **1 Basic Folder**.
3. Click **Create**.
4. Double-click on the name for **Folder 1**.
5. In the track name window, type `Stereo Tracks`.
6. Click the **Next** button.

7. In the track name window for **Folder 2**, type `Mono Tracks`.

8. Click **Next** or press *Command* + right arrow (*Ctrl* + right arrow on Windows) to move to the next track.

9. For Folder 3, type in `Other Tracks`.

10. Click **OK**.

11. Either in the **Tracks List** area or in the **Edit** window, *Shift* or *Command* (*Ctrl* on Windows) and click all the stereo audio tracks, and drag them onto the `Stereo Tracks` folder track.

12. Do the same for the mono audio tracks, placing them in the `Mono Tracks` folder track and the other tracks in the **Other Tracks** folder.

13. Click on the first stereo audio track under the **Stereo Tracks** folder and click on the output dropdown in the **I/O** window.

14. Select **track | Stereo Tracks** (see *Figure 1.12*):

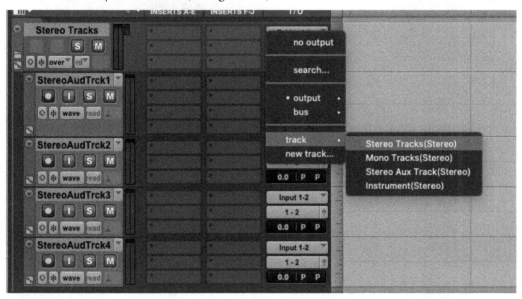

Figure 1.12: Setting the track output to a routing folder

15. Multi-select with *Command* (*Ctrl* on Windows) or *Shift* the other three stereo audio tracks.

16. Hold *Option/Alt* + *Shift* on your keyboard and set the output of one of the stereo audio tracks to the same **Stereo Tracks** folder.

17. Do the same for the mono audio tracks.

How it works...

Folder tracks do not automatically organize tracks under them. Once created, you need to move other tracks into them. Additionally, when creating routing folders, the audio outputs from the tracks placed within them still need to be set to the routing folder itself. Basic folders have no routing and are simply there for organization.

There's more...

Within the **Edit** window, you will see a tiny folder icon on the left-hand side of the **Track Header** area (see *Figure 1.13*). You can click that to collapse and expand the folder, hiding and showing the tracks within it. In the **Tracks** list, you will see that the folder tracks have a tiny triangle that you can click on to achieve the same effect. Routing folders mix down the audio routed to them. Any volume changes or effects added to these tracks will affect the summed signal, just like an Aux track:

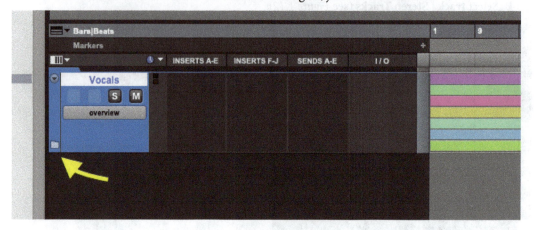

Figure 1.13: The Folder Expand/Contract button

Grouping tracks for editing, mixing, and viewing

So far, we've explored how to use Aux tracks and folder tracks to set up our session. Another powerful tool that can be used for editing, mixing, and viewing tracks is groups. By default, you already have one group for all your tracks named <ALL |. We'll create some new groups and see how they function.

Getting ready

This recipe requires a Pro Tools session with four stereo tracks, four mono tracks, and three Aux tracks. Your session should also have a left-hand side drawer with **Track List** and the **Groups** list visible. If you don't see it, click the tiny arrow in the bottom left of the **Edit** window (see *Figure 1.14*) or go to the menu bar and select **View** | **Other Displays** | **Track List** (see *Figure 1.15*):

Figure 1.14: The Track sidebar's Show/Hide button

Figure 1.15: Enabling Track List to be viewable

How to do it...

Now, let's create Groups and assign different bundles of tracks to each group:

1. *Shift* or *Command* (*Ctrl* on Windows) and click all the stereo tracks in your session.
2. Go to the **Groups** list at the bottom left of the **Edit** window and click the arrow on the top right of it.
3. Select **New Group...** (*Command* + *G* on macOS, *Ctrl* + *G* on Windows).
4. Under the **Name** field, enter Stereo Tracks; everything else can be left as is. Then, click the **OK** button.
5. With no tracks highlighted, click the triangle at the top right of the **Groups** list and select **New Group...** again.
6. Name this group Mono Tracks.
7. Under the **Tracks** tab, *Shift* or *Command* (*Ctrl* on Windows) and click all the mono tracks from the **Available** area, and click the **Add** button to have them move to the **Currently in Group** area.
8. Click the **OK** button.
9. Create one more group for the Aux tracks in the session.

How it works...

Groups don't do anything until they are active. You can click on any group within the list to activate it (it will be highlighted in blue), and now any change in the mix level or edit you apply to one track within the group will be applied to all of them. If you only want to manipulate the edit or mix portion of a group, you can change that within the **Groups** list by right-clicking on a group and selecting **Modify**. This can also be used to add or remove tracks from a group.

Groups can be useful for many scenarios, especially when editing. If you have a sound effect that's made up of many layers of sound, or music in multiple tracks that you want to ensure stays in sync, grouping them is the easiest way to do so.

There's more...

Notice the **az** box at the top right of the **Groups** list? (See *Figure 1.16*.) This is called **Groups Keyboard Focus**. Click on it to make it active for the **Groups** list; now, you activate or deactivate those groups by simply typing in the letter next to the group on your keyboard:

Figure 1.16: Groups Keyboard Focus

What I find super helpful with groups is the ability to easily hide and show groups from your session by holding *Control* and clicking on them. Try it out by *Control* and clicking on one group; you will notice it hides the other groups. If you want to have more than one group appear, you can hold *Shift + Control* while clicking on them.

A track can be added to multiple groups, so with the aforementioned tip in mind, you can create a range of viewing configurations for your project. If you want to only focus on dialog and voice-over, you can have a group for that. If you want all the tracks except vocals to appear on the screen during a song, you can build a group for that too. To show all tracks, simply *Control* and click on the **<ALL>** group.

Using memory locations within a project

Memory locations are a great way to communicate ideas and instructions between team members. They are essentially markers that can be placed on the timeline. They can also be one method of putting notes for yourself or others. I also use markers extensively for motion picture and linear media to denote scene changes since so much sonic information changes between scenes. In this recipe, we'll add some simple markers to our project in a few different ways and show how they can be used.

Getting ready

For this recipe, a blank Pro Tools session is all you need. Make sure that the **Markers** ruler is visible in the **Edit** window. You can do this by going to the menu bar and selecting **View | Rulers** and confirming **Markers** is checked.

How to do it...

We'll now create some memory marker locations. Follow these steps:

1. In the **Edit** window, move your mouse to anywhere on top of the **Markers** ruler.

2. Hold *Control* (*Start* on Windows) and click anywhere on the ruler.

3. Name the memory location Marker 1.

4. Click the **OK** button.

5. Press the spacebar or the **Play** button in the transport window to start playback and progress the timeline.

6. As the timeline indicator is moving, press *Enter* (not *Return*) on your keyboard to add markers.

7. Click and drag markers to move them around and relocate them.

8. Double-click on a memory location to bring up its properties and edit them.

9. Hold *Option* (*Alt* on Windows) and click on a marker to remove it.

10. Click on a marker to have the timeline insertion indicator snap to it.

11. Go to the menu bar and select **Window | Memory Locations**.

12. Click on a marker in the list to snap the timeline insertion indicator and move to it.

13. *Option* (*Alt* on Windows) and clicking on a marker in the memory location window also removes it.

There's more...

Memory locations can also hold other information, as seen in the **Memory Locations** properties window. You can have a selection saved in memory if it's an area you need to comment on or make notes about. You can also save **Zoom**, **Track Visibility**, and other parameters. With these types of markers, you can set the time properties to **none** so that it doesn't change where on the timeline your insertion goes.

Markers can be recalled with the number pad by typing the period key (.) followed by the marker number and closing with another period. So, going to marker 12 would require typing in [.12.].

Markers can also be exported from the **Memory Locations** window to a text file. Click on the triangle at the top right of the window and select **Export markers as text...** (see *Figure 1.17*). You will see several options in terms of how to treat the text file. This can be very useful for spotting sessions or planning and preparing sessions. There are third-party tools such as **EdiMarker** from **SoundInSync** that can import text files into markers as well:

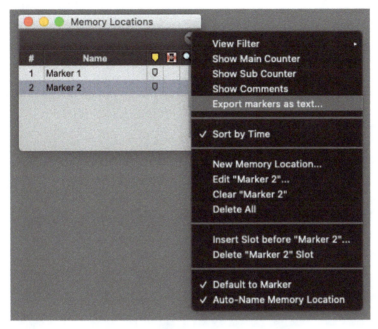

Figure 1.17: Export markers as text…

2

Importing and Organizing Audio

Before you can start working with audio, you will need to import the media you want to work with into your Pro Tools session. Pro Tools offers several ways to bring in audio, and then organize them for better collaboration with others. In this chapter, you'll learn about those methods, but also their impact on the project files and sessions. We'll also discover how to import data from other file formats and sessions, and we'll look at other tools that can assist with file management and organization.

This chapter will cover the following topics:

- Importing audio into a session with the Import Audio window
- Dragging audio directly into a session
- Importing audio from an AAF/OMF or another session
- Relinking broken audio
- Using the Clip List area for audio management
- Collaborating with others using session data

Technical requirements

The recipes in this chapter will require at least Pro Tools Artist, preferably Pro Tools Standard. The example sessions and audio files for each recipe can be found at `https://github.com/PacktPublishing/The-Pro-Tools-2023-Post-Audio-Cookbook/`.

Importing audio into a session with the Import Audio window

While there are several ways to import audio into Pro Tools, the method that provides the most control and options is through the **File** menu. Often, issues that arise from Pro Tools sessions are due to audio not being imported correctly; mastering this menu and its options will cause less frustration in the long run.

Getting ready

You will need an empty Pro Tools session and some audio files to import into it for this recipe.

How to do it...

For this recipe, we'll import multiple files from different sources into Pro Tools. We'll see what the different behaviors are, and how it affects the workflow:

1. Go to the menu bar and select **File | Import | Audio….**

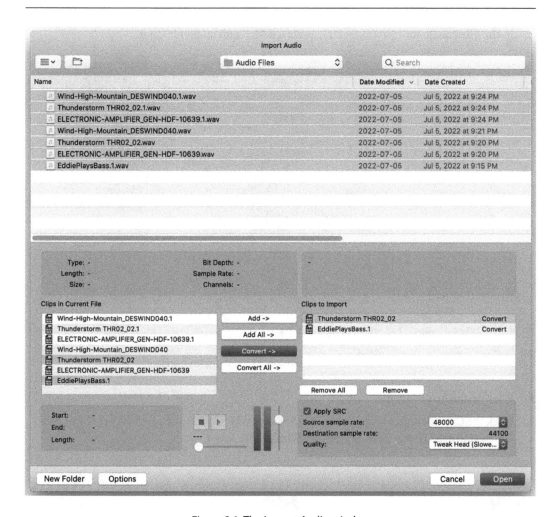

Figure 2.1: The Import Audio window

2. In the **File Navigator** (see *Figure 2.2*), locate a folder with audio files in it.

Figure 2.2: The Import Audio window's File Navigator (top half)

3. Click on the first audio file.

4. Click the **Copy** button (**Copy Files** in Windows) (see *Figure 2.3*) – if you don't see **Copy**, click **Convert** instead (**Copy Files** or **Convert Files** in Windows).

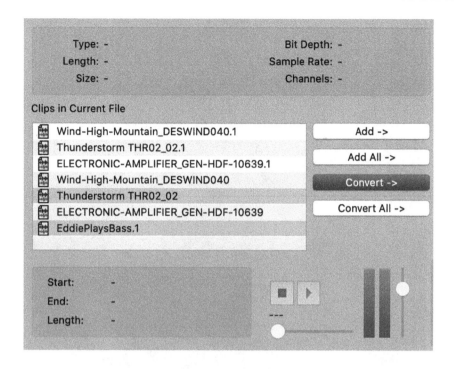

Figure 2.3: The Import Audio window's clip preview (bottom left)

5. Click the next audio file in the list.

6. Click the **Convert** button.

7. Click the next audio file in the list.

8. Click the **Add** button (**Add Files** in Windows).

9. Click the **Open** button (**Done** in Windows).

10. When prompted to **Choose a destination folder**, click **Open** (**Use current folder** in Windows).

11. In the **Import Audio** window, select **Clip List**:

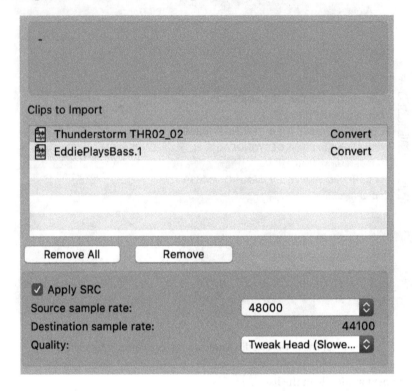

Figure 2.4: The Import Audio window's import lift (bottom right)

How it works...

The audio import options might seem straightforward, but there are some nuances to consider that can greatly affect your project. The first thing you may have noticed is the fact that the **Import Audio** dialog lists **Clips in Current File**. Pro Tools can work with **Clip Groups**, which essentially are multiple clips within a file. While it can still be useful in modern audio production workflows, I rarely use this feature. It is still useful to be able to corral audio files from different sources and import them in essentially one step. Once a clip has been added to the **Clips to Import** area, you can navigate to other parts of your filesystem and find other audio files to import. Here are some notes about the different behaviors.

Copy

This will make a copy of the audio file(s) you've selected in the destination of your choice. By default, this is the Audio Files folder within your Pro Tools Session folder.

Convert

You will notice that the **Copy** button changes to **Convert** when there is a sample rate mismatch. As-s, the file should be converted to maintain the correct playback speed. You can see the options for **Sample Rate Conversion** (**SRC**) in the **Apply SRC** area of the import options. Pro Tools should automatically detect and select the correct settings to apply, but you could always choose alternative options if you want to have some fun with playback speed.

Add

The **Add** function will not copy the audio file into a destination folder; instead, it will leave it in the place you've selected it from and simply reference that location when playing back the clip. This can be advantageous if you prefer to not duplicate files or divide a project into multiple Pro Tools files and want to have them all reference the same source audio files. However, take caution if your audio files move, as this will break the reference and the files will show as missing. Also, if you attempt to add a file that is not the same sample rate as your project, Pro Tools will warn you that the playback speed will not sound correct.

Destination folder

As mentioned previously, all files that are copied or converted will be sent to a folder of your choosing. By default, this is the **Audio Files** folder within your Pro Tools `Session` folder, but you can select a different destination if you wish. You could, for instance, choose to create subfolders within your **Audio Files** folder for different categories of sounds. If you know that a file will be used by another project, you could also opt to change the destination to that project's folder as well.

Clip List versus New Tracks

After the **Processing Audio** dialog is complete, you will get the **Audio Import Options** window. From here, you can choose to place these clips into **New Tracks** or the **Clip List** area. If you are receiving files from another session as stems or printed tracks that are all lined up, then **New Tracks** makes more sense. You have further options to place the audio clips at **Session Start**, which is the beginning of the session; **Selection**, the current selection in the timeline; or **Spot**, which opens another dialog window with specific timing options on where to place the clips.

The **Clip List** area is the best place to import audio clips that will be placed into another track within your session. If you don't see the **Clip List** area after importing, click the small arrow in the bottom right of the **Edit** window (see *Figure 2.5*), or go to the menu bar and select **View | Other Displays | Clip List**:

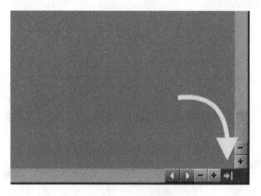

Figure 2.5: The Clip List drawer's Open/Close button

Dragging audio directly into a session

Depending on the scenario and how your files are organized, it's sometimes advantageous to drag files directly into a Pro Tools session. Audio clips and tracks behave differently than using the **Audio Import** function, so it's important to know the difference.

Getting ready

For this recipe, you will need a Pro Tools session and some audio files. To fully appreciate all the nuances, your session should have both mono and stereo tracks, and you should have both mono and stereo audio files.

How to do it...

In this recipe, we'll drag several different audio files into a session in different ways and see how they behave. To do this, follow these steps:

1. Navigate to a mono audio file in your filesystem and drag it into a mono track.
2. Try dragging the same mono audio file into a stereo track.
3. Find a stereo file and drag it into a stereo track.
4. Try dragging the same stereo file into a mono track.
5. Drag multiple audio files into the clip list.
6. Drag multiple audio files into an empty area under the existing tracks.

How it works...

Dragging audio files into Pro Tools behaves differently, depending on the context. Let's break down the different scenarios we played with previously.

Dragging in mono files

Mono audio files only have one audio channel. Attempting to drag a mono audio file into a mono track provides no issues as the channel counts are consistent. However, you will notice that you cannot do the same with stereo tracks. You will not see any clip audio at all, which is Pro Tools indicating to you that this is not possible.

Dragging in stereo files

Stereo files behave similarly to mono files in that it's expected that a stereo clip is added to a stereo file. Unlike mono files, however, you can add a stereo file to mono tracks, if there are at least two consecutive mono tracks in your session. You will see that the left and right channels are separated into separate mono clips. This can be beneficial if you are working with audio files where the discrete audio channels of a stereo file are not stereo images – for example, if you're working with audio that was recorded on set for a motion picture and the boom microphone was recorded to one channel, and the lavaliers to another.

Sample rates matter

When ingesting audio into your session through the **Audio Import** dialog, you can decide whether an audio clip is added or copied to a destination folder. When dragging audio files into a session directly, Pro Tools will automatically **add** the file (as in reference it in its original location) if the sample rate of the file matches that of the session. If you drag in a file that has a different sample rate, Pro Tools will automatically apply sample rate conversion to match that of the session and save the converted file in the `Audio File` folder in the project's session folder.

Creating tracks as you drag

If you need a new track to place audio clips into, instead of creating the tracks first, you can drag them directly into a session. This is similar to using the import dialog and selecting **New Track** when prompted for **Audio Import** options. When using this method, the track name will be set to the filename, and its channel count will match the source file. This is a good method for importing stems from another project or music tracks that are already synchronized from the start of the file.

There's more...

These methods and nuances are not exclusive to dragging audio files directly into Pro Tools. The same behavior can be expected when dragging clips into the session from the **Clip List** area. Knowing what to expect when bringing media into a track will prevent confusion when channel counts and sample rates don't match.

Importing audio from an AAF/OMF or another session

Individual audio clips are great when you are creating a session from scratch, but much of the professional audio workflow expects collaboration between departments and delivery from one phase of production to the next. As such, you should expect to be handed audio session data either in special audio formats such as OMF or AAF or from another Pro Tools session entirely. Knowing what to expect when importing audio from one of these files is important if you want to succeed professionally.

Getting ready

This recipe requires either an AAF, OMF, or another Pro Tools session with audio data. Encapsulated files or files with separate audio are both acceptable. An example OMF has been provided on the GitHub page for this book: `https://github.com/PacktPublishing/The-Pro-Tools-2023-Post-Audio-Cookbook/`.

How to do it...

We are now going to import an OMF file from a short audio drama that has different types of audio tracks into a Pro Tools session. Follow these steps:

1. Go to the menu and select **File | Import | Session Data...**.

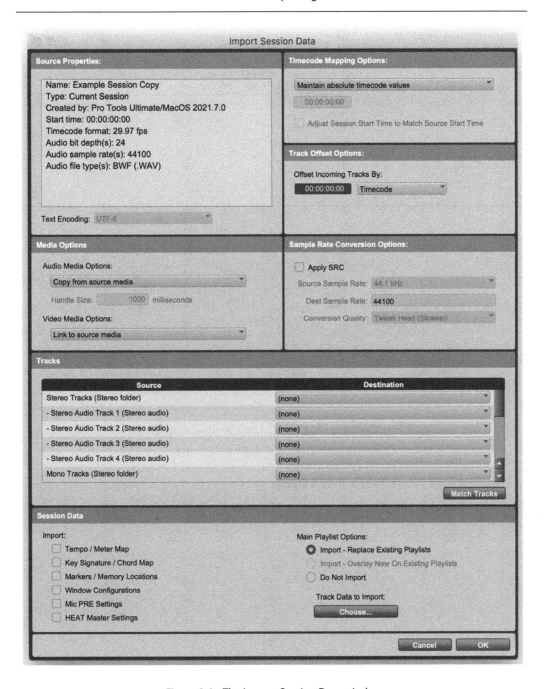

Figure 2.6a: The Import Session Data window

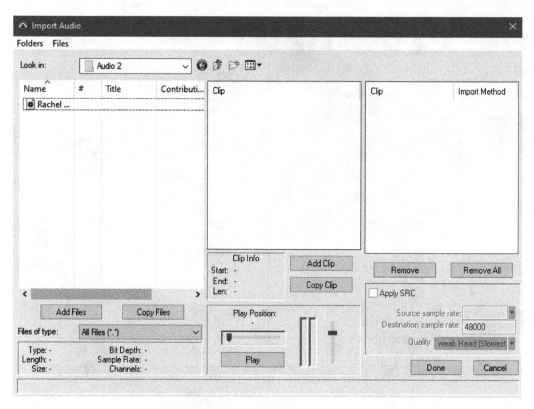

Figure 2.6b: The Import Audio dialog on a Windows system has
a different layout, but functions the same way

2. Find and select the OMF file and click **Open**.

3. Under **Timecode Mapping Options**, select **Maintain absolute timecode values**.

4.   Under **Media Options**, set **Audio Media Options** to **Copy from source media**.

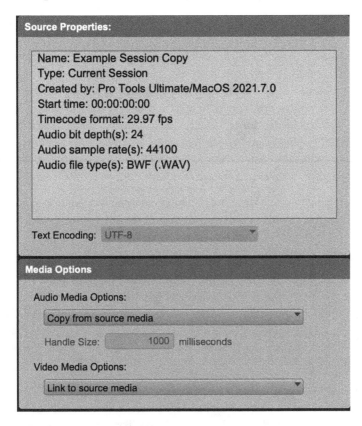

Source Properties:

Name: Example Session Copy
Type: Current Session
Created by: Pro Tools Ultimate/MacOS 2021.7.0
Start time: 00:00:00:00
Timecode format: 29.97 fps
Audio bit depth(s): 24
Audio sample rate(s): 44100
Audio file type(s): BWF (.WAV)

Text Encoding: UTF-8

Media Options

Audio Media Options:

Copy from source media

Handle Size: 1000   milliseconds

Video Media Options:

Link to source media

Figure 2.7: The Import Session Data window's Source Properties and Media Options (top left)

5.   Under **Tracks**, make sure all sources are highlighted and **Destination** is set to **New Track**.

6.   Under **Session Data**, check **Import Clip Gain** and **Import Volume Automation**, and uncheck the other options.

7.   Under **Main Playlist Options**, select **Import – Replace Exiting Playlists**.

8.   Click **OK.**

How it works...

OMFs and AAFs are special files that are designed to allow media to be shared across different software applications. They both have different standards in terms of how they are created and read by the software, but at their core, they are text files that list all the clips used in a session, which track they go on, when they start, and when they stop. Both files can include media encapsulated within the file, but for larger sessions, having the media separate tends to cause fewer problems in the long run. Clips can also be trimmed so that only the audio used in the session is exported, which "handles" a little bit of extra audio both before and after the cut points to allow some flexibility. I typically ask for 3 seconds of handles, but I've seen some prefer more. AAF also allows for what is essentially only a text file and does not re-export the media. This can help you save on file space, but can lead to broken links when importing if file structures are changed (the next recipe goes over relinking broken audio).

There are a lot of different options to go over in the **Import Session Data** window, so let's go over them.

Source Properties (Figure 2.7)

This is a simple text readout of the project you are importing. This is a good way to confirm that the project was delivered correctly and that the settings are accurate for what you are expecting. If you notice something is not correct, you can request changes from whoever delivered you the files, or you can change your project's session settings to match. If it is a **Sample Rate** mismatch, you will either need to apply sample rate conversion or create a new session.

Media Options (Figure 2.7)

It's usually a safe bet to copy the audio files from the source media, especially if the audio files are embedded or encapsulated within the OMF or AAF file. You can, however, choose to link to source media where possible or consolidate the files. Linking will work like adding files to a project and leaving the files in their original location by simply referencing them. Consolidating will trim the files to the size used within the clips embedded. You can set the number of handles in milliseconds that will be left, which is how much outside of the in and out points will be trimmed, allowing you to have some flexibility. Only in extreme cases in which copying all the data would be extremely taxing from a data standpoint would I opt for consolidating media, although in some scenarios, I have seen excessive amounts of audio data provided in a session, and consolidating made more sense.

Timecode Mapping Options (Figure 2.8)

If everything was done correctly, you will usually opt to maintain the absolute timecode values. However, in some scenarios, what was delivered is incorrect.

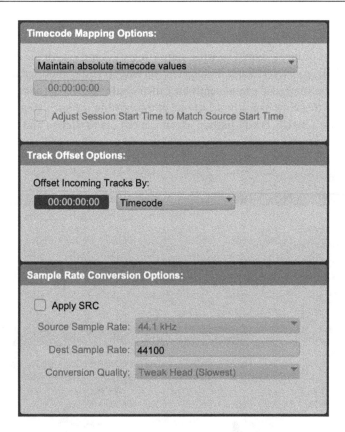

Figure 2.8: The Import Session Data window's Timecode Mapping Options,
Track Offset Options, and Sample Rate Conversion Options (top right)

For example, I request that the picture start begins at 01:00:00:00, which is standard in motion picture, but if the editor or production team is not familiar with this, they may simply have it start at 00:00:00:00, which is the default in most non-linear editing software. I can either attempt to educate them on how to properly set up their timeline before delivery – and in the case with clients where I wish there to be an ongoing relationship, I usually will take the extra effort – or I can simply remap their session data to the time I wish to use. Maintaining relative timecode values means that whatever your session start is will be remapped to the source files, or you can simply map the start time of their session to a specific value. Be warned that if you intend to send your audio project to another department or person, they should be aware of this.

Sample Rate Conversion Options (Figure 2.8)

If the source audio's sample rate does not match the sessions, you can make some decisions in terms of how the sample rate conversion is handled. These are the same as the options that are available in the **Audio Import** dialog window (see the previous chapter for more details).

Tracks (Figure 2.9)

This is where most of your focus should be when importing session data. All the audio tracks from the source are listed here. Any track that is highlighted will have its audio imported into your session, whereas deselected tracks won't, and consequently have their destination set to **(none)**. You can choose to have tracks mapped to a **New Track** or select one of the existing tracks in your session from the dropdown. There is also a **Match Tracks** button, which you can use to attempt and automatically line up similarly spelled tracks.

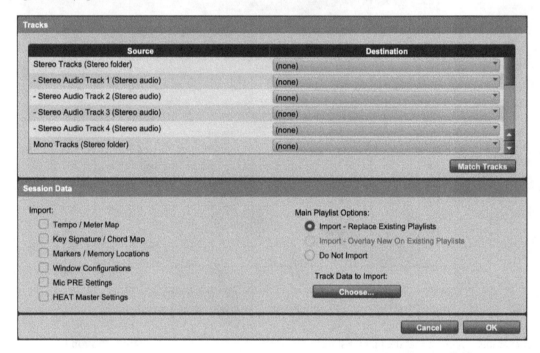

Figure 2.9: The Import Session Data window's Tracks and Session Data (bottom half)

Be warned that the default behavior for matching a track is to overwrite and replace the data in that existing track. If you wish to retain that data, either set the source track to a **New Track** as its destination and manually move the audio after, or you can change **Main Playlist Options** to **Import – Overlay New on Existing Playlists**. This will retain existing clips in your track and insert new audio around them.

Session Data (Figure 2.9)

Besides Main **Playlist Options**, there are also several other checkboxes to look over in this section. These can change according to the type of file you are importing, but here are some common ones you might see:

- **Import Rendered Audio Effects**: This copies audio effects rendered over clips if selected or discards them and uses the original audio if unchecked (only available when importing from software that supports this option).

- **Import Clip Gain**: This retains the clip gain set in the previous project if checked or discards it if unchecked (again, not all software supports this).

- **Import Volume Automation**: This retains the volume automation as set when it was exported. Unchecking this will remove all automation and set the volume to 0 dB.

- **Only Include Clips on Timeline**: If the exported session allows for project data within the clip list or bin to be exported, then this will ignore those files and only bring in files that are on the timeline.

- **Pan odd tracks left/even tracks right**: This can be helpful if you are working with stereo tracks, as OMFs can only import mono tracks. By default, stereo tracks that are imported will be split into two tracks named **TrackName L** and **TrackName R**. Having them panned left/right makes sure they are panned correctly. This can confuse things if you're importing a combination of mono and stereo tracks, though.

- **Markers to Import**: This allows you to select which (if any) markers you want to import.

- **Track Data to Import**: This allows you to get granular with what gets imported into a track – for example, whether you can import only plugin information or only input/output configuration. This can be useful for situations where you want to bring in specific data from a session, but not all the audio files.

With all these options, importing data from another session can be very flexible. For podcasts that use similar mic and audio setups routinely, you can set the source tracks to tracks in a template session with effects and routing already set and import only **Clips and Media** using the **Track Data to Import** options. This will leave all the other hard work intact and bring in only the audio. You can also import only session markers if you want to bring in notes from another collaborator.

There's more...

Some DAWs allow audio clips with different channel counts to exist on the same timeline. When exporting OMF or AAF files, these clips should automatically be separated into different tracks. For example, you could have three tracks named **Effects**, **Left Effects**, and **Right and Effects Mono** to represent those different channels.

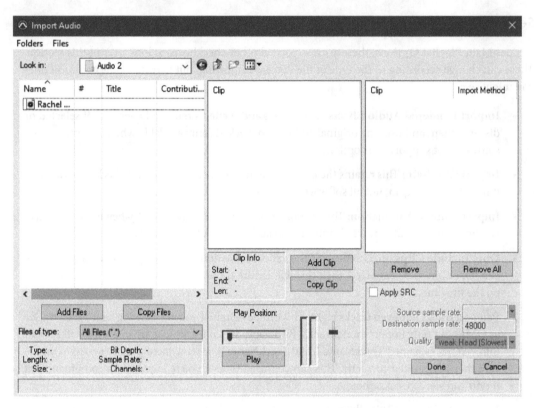

Figure 2.10: The Import Audio dialog on a Windows system has
a different layout, but functions the same way

While rare, it is also possible for an issue to occur where tracks are not separated, and tracks try to contain both types of clips with mismatched audio channels. Pro Tools is not able to import these tracks correctly and will simply crash instead. If you have issues importing tracks into Pro Tools, the best method of solving the issue is to attempt importing only a few tracks at a time to isolate which tracks are causing the problem. If you can, you can ask the sender of the file to examine the tracks and see if there are channel mismatches or other possible causes of failure. It may also be possible to import the session data into another DAW to identify and rectify the issue.

Relinking broken audio

It's a sad truth that everyone at some point or another will face this issue. Since Pro Tools references audio files in a specified location, any change in the filename or location can break the link, causing it to appear offline in the session. Pro Tools also uses a unique identifier for ensuring that similarly named files do not get mixed up. This can cause problems from time to time, especially with OMFs and AAF files. In most scenarios, Pro Tools will try to relink the audio when it detects that some files are missing, but you can also manually relink them. We'll go over how to do that in this recipe.

Getting ready

For this recipe, you will need a Pro Tools session with some audio placed into tracks. The audio files will need to have their links broken. You can do this by relocating the audio files into a subfolder within the **Audio Files** folder. Make sure you do this with the Pro Tools session closed; otherwise, performing such a change while it's open can crash Pro Tools. When you open a session with broken links, a warning will appear, asking you what you want to do with **Missing Files**. Select the **Skip All** option when prompted:

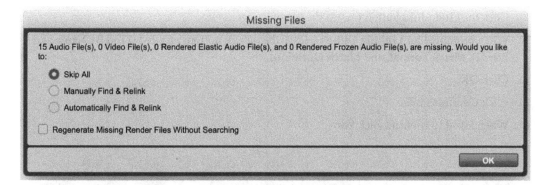

Figure 2.11: The Missing Files window

How to do it...

We'll be using the workspace tool to manually relink some broken audio links for this recipe. You can do this by following these steps:

1. In the menu, go to **Window | New Workspace | Default**.

2. In the sidebar on the left (the area labeled **Locations**), click on the icon that specifies your session name.

3. Double-click the folder named **Session Audio Files** (**Audio Files** on Windows).

4. Locate audio files that are broken – they will be grayed out and in *italics* (see *Figure 2.12*):

▼ ⬜ Session Audio Files	Catalog	
ADL2766381077L-2102-224	○ *Audio File*	▶
ADL2766381077L-2743-300	○ *Audio File*	▶
ADL2766381077R-2102-224	○ *Audio File*	▶
ADL2766381077R-2743-300	○ *Audio File*	▶

Figure 2.12: The Workspace window showing missing files

5. *Shift* or *Command* and click (*Ctrl* on Windows) to highlight them.

6. Right-click (or *Control* and click on macOS) on any of the highlighted files and select **Relink**.

7. In the **Relink** window, use the **Select Areas to Search:** area and navigate to your session's audio files folder on your disk (this could be an internal or external hard drive or SSD; do not navigate to the session itself, which is denoted by a Pro Tools icon).

8. Click the checkbox next to the `Audio Files` folder.

9. Select all the files within the **Search Files to Relink:** area.

10. Click the **Find Links** button toward the top of the window.

11. When prompted for **Linking Options,** leave the default options enabled (**Find By Name and File ID**, **Match Format**, and **Match Duration**).

12. Click **OK**.

13. Click **Commit Links**.

14. When asked to confirm, click **Yes**:

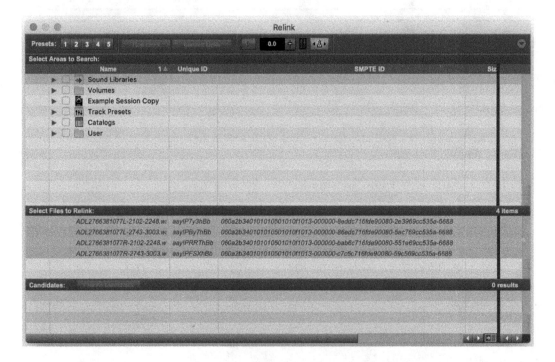

Figure 2.13: The Relink window

How it works...

Workspaces function like a file explorer and manager within Pro Tools. Tools such as **Sound Libraries**, **Track Presets**, and **Catalogs** can all be accessed here, but our focus is on the files within the open session. Examining the files listed here gives you a lot of information regarding the audio imported into your session and allows you to see which files are unlinked – the text is grayed out and in italics. You can then use the **Relink** command to open the relink window.

Within the **Relink** window, you can **Find Links** for specific files and in a specified location. You can also click on a higher-level folder or even entire volumes if you are unsure where the files could be located, but this takes significantly more time than locating the correct folder. You should always be cautious when moving files around or renaming files or folders with Pro Tools sessions, but it's good to know where to solve linking errors if they come up.

There's more...

As mentioned at the beginning of this recipe, it is sometimes possible to have linking errors appear when importing session data. If this is the case, Pro Tools might not recognize the audio files correctly, and it will remain this way unless you choose less restrictive forms of searching for files. After clicking the **Find Links** button, the **Linking Options** window will appear with the default options of finding **File by Name and File ID**, as well as **Match Format** and **Match Duration** selected. You can use **Find by Name** and deselect the other options to widen your search. Pro Tools will warn you about the dangers of linking a file that does not match the duration, so make sure you do link the correct file. This also reinforces the importance of good file naming.

You can also try the **Find All Candidates** button at the bottom. Instead of finding just one specific file, Pro Tools will offer you one or more files within the **Candidates** area of the window. Click the empty button to the left of the filename to see a **Link** icon appear – this denotes that this file will be used when relinking.

Using the Clip List area for audio management

The **Clip List** area is where all the audio files imported into a Pro Tools session are listed, whether they are used or not. Unlike other editing software, Pro Tools cannot organize files into folders within the **Clip List** area. This can be very daunting to use, but there are some tools available for users to get the most out of the **Clip List** area.

Getting ready

This recipe expects you to have a Pro Tools session with multiple audio files imported into it. Your session should also have the right-side drawer with the **Clips List** area visible. If you don't see it, click the tiny arrow in the bottom right of the **Edit** window or go to the menu bar and select **View | Other Displays | Clip List** (see *Figure 2.14*):

Figure 2.14: Enabling Clip List

How to do it...

This recipe will go over some of the commands you can use with **Clip List** to find audio clips and manage them. Follow these steps:

1. In the **Clip List** area, click the triangle in the top right.

2. Select **Find…**.

3. Start typing in some letters that match one of the filenames and observe how the **Clip List** area changes.

4. Click the triangle to the right of the **Name** field and select **Insert Entry**.

5. Type in another set of letters different from the first into the **Name** field.

6. Click the triangle again and select **Insert Entry** once more.

7. Click the triangle and toggle between the two entries.

8. Select one of the entries and click **OK**.

9. Back in the **Clip List** area, click the triangle at the top right and select **Clear Find.**

10. Click the keyboard focus (the small **az** button at the top right).

11. Type in a filename.

12. Hold *Option* (*Alt* on Windows) and click on a file *(audio warning – make sure your monitors/ headphones are not set loud).*

How it works...

As mentioned previously, the **Clip List** area is one large collection of all the audio clips and files in your session. This can make finding clips very frustrating if you don't use the Find commands within the **Clip List** area. Find commands can also be hard to narrow down, but the fact that it updates the contents of the list live as you type in letters can help narrow things down, and using the entries to help sort things is another often used technique. Keyboard focus is also very helpful in that you can simply type in the filename, but that also removes the quick command functionality from the **Edit** window.

There's more...

If you want to keep the keyboard focus on the **Clip List** area but have quick commands available for edit functions, simply hold *Control* (the *Windows* key on Windows) in combination with the keyboard letter to have that command work in the **Edit** window. For example, the letters *r* and *t* on your keyboard zoom out and in respectively when the keyboard focus is on the **Edit** window. If the keyboard focus is on the **Clip List** area, you can press *Control* + *r* and *Control* + *t* (the *Windows* key + *r* and the *Windows* key + *t* on Windows) for the same effect.

See also

There are many third-party tools available that provide extra functionality for organizing and finding media for your projects. See the *Appendix* for a list of them.

Collaborating with others using session data

While working alone is perfectly fine for many projects, most professional environments are collaborative. Being able to send projects and session data between collaborators with as little friction and issues as possible will prevent headaches and frustration. In this recipe, we'll go over some best practices for saving your projects as standalone sessions and sending the data to other projects. Some workflows also benefit from having multiple separate sessions for different areas of work, so knowing how to save your session data to send to yourself for later is useful.

Getting ready

For this recipe, you will need a Pro Tools session open with audio clips imported and placed into tracks.

How to do it...

We're going to take a Pro Tools session and save a copy to be imported into another session. Follow along with these steps:

1. Go to **File | Save Copy In....**

2. For **Format**, select **Session (Latest)**.

3. Leave **Session Parameters** at their defaults (the same as the original session).

4. For **Items to Copy**, enable **Audio Files**.

5. Click **OK**.

6. In the **Save** window, navigate to a location you'd like to save the session in and give it a name.

7. Click **Save**:

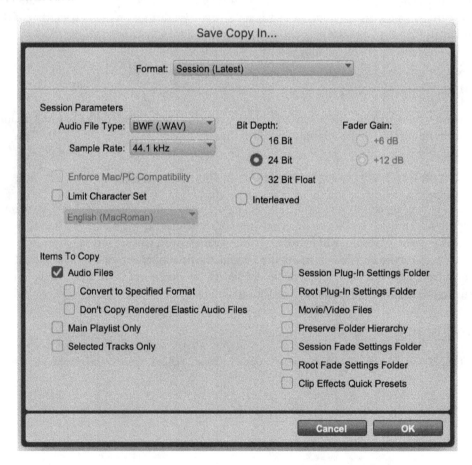

Figure 2.15: The Save Copy In… window

How it works...

What we just did was save a copy of our session with all the audio files copied into its `Audio Files` folder. This is important as a way of collaborating between sessions since Pro Tools can either copy audio files into a project session's `Audio Files` folder or simply reference the files added to it from another location. When using the **Save Copy In...** command and selecting the audio files to copy, it will ensure that all files present in the **Clip List** area are copied to the new session's `Audio Files` folder. If you want to send this project somewhere else, make sure you include the entire folder, not just the Pro Tools session file (`.ptx`).

There's more...

If you are sending a project to another system and need to have it in a specified format, the **Save Copy In...** command can do this for you. The **Format** option at the top of the window allows you to select an older version of Pro Tools from a dropdown (from versions 5.9-6.9 or 7-9). This will typically export a `.ptf` file, along with the other folders. These files can be opened in newer versions of Pro Tools, but newer files (such as `.ptx`) cannot be opened in older versions of Pro Tools. While you may think it wouldn't be common to have to open in older versions of Pro Tools, the reality is that due to Avid's business practices, many studios opt to "freeze" their software at a particular version. Most versions of Pro Tools are designed to work with specific operating systems and, by extension, certain hardware. If a large studio invests a large amount of money into high-end hardware, it would be impractical to refresh all that gear just to have it work with newer versions of Pro Tools.

The other options available also allow you to select a target **Sample Rate** and **Bit Depth**, and you can further narrow down the items you are saving by choosing to export only **Selected Tracks** or only **Main Playlists**.

One other important note about saving projects in this method is it also saves markers. If you have notes you want to provide to collaborators, you could create a session with just markers exported; then, those can be imported into another project.

3
Faster Editing Techniques

When beginning to use Pro Tools, the interface and UI can be challenging. Even after years of working with it, I often find trying to interact with the software cumbersome. The good news is with time, practice, and patience, you can find little ways to improve your editing techniques that speed things up considerably when used together. In this chapter, you will be shown some different techniques for editing clips within Pro Tools that will provide some insights into how to most effectively use the software. We'll explore which tools are best suited for certain functions, and how to navigate Pro Tools using the keyboard.

This chapter covers the following topics:

- Using the **Smart Tool** for quick edits
- Editing with the keyboard
- Generating and manipulating fades
- Using **Edit** modes correctly
- When to use the **Pencil Tool**
- Copying and pasting clips
- Printing effects on clips using AudioSuite

Technical requirements

This chapter requires at least Pro Tools Intro.

Using the Smart Tool for quick edits

Getting to know the **Tool Bar** and all its functions is important for anyone working with Pro Tools. However, with the introduction of the **Smart Tool**, the vast majority of your work can be completed with one context-sensitive tool. The tool's actions change according to where it's positioned, and knowing how and when those different actions appear—and their effect—will help you edit faster and more effectively.

Getting ready

This recipe requires a Pro Tools session with at least one audio clip placed into a track.

How to do it...

For this recipe, we'll explore all the different ways a clip can be manipulated using the Smart Tool. Follow along with these steps:

1. In the **Tool Bar**, click on the bar above **Trim Tool**, **Select Tool**, and **Grabber Tool** to activate the **Smart Tool**:

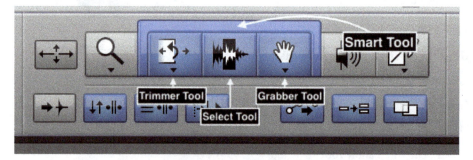

Figure 3.1: The Smart Tool

2. Position the cursor over the middle of a clip.

3. Move the cursor to the bottom half of the clip to have it change to the **Grabber Tool** and click on the clip (see *Figure 3.2*).

4. Move the cursor to the top half of the clip to have it change to the **Select Tool** and click on the clip (see *Figure 3.2*):

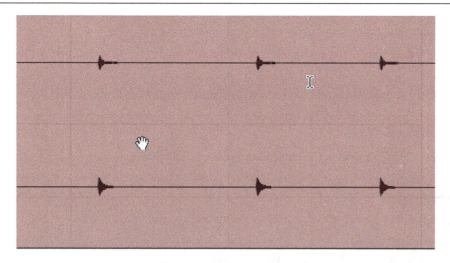

Figure 3.2: On the left is the Grabber Tool (bottom half of the clip),
and on the right is the Select Tool (top half of the clip)

5.   Press *command + E* (*Ctrl + E* on Windows) on your keyboard to add an edit to the clip; you can also use the *B* key with the keyboard focus on the **Edit** window.

6.   Position the cursor over the bottom of the edit just created to have it change to the **Crossfade Tool**:

Figure 3.3: The Crossfade Tool

7.   Click and drag the **Crossfade Tool**, starting from the edit point to either the left or right.

8.   Move the cursor to the bottom center of the crossfade to have it change to the **Grabber Tool**.

9.   Click and drag the crossfade to the left or right to reposition it.

10. Position the cursor over the middle center of the crossfade to have it change to the **Crossfade Tool** again.

11. Click and drag the fade and observe how the shape of the fade changes.

12. Position the cursor toward the inside edge of the crossfade in its upper half to have it change to the **Trimmer Tool**.

13. Click and drag the edge of the crossfade and observe its behavior.

14. Move the cursor to the outside edge of the crossfade and notice the direction of the **Trimmer Tool** change.

15. Click and drag the outside edge of the crossfade and observe its behavior.

16. Navigate to either the beginning or end of the clip and position the cursor toward the bottom half of the clip to have it change to the **Trimmer Tool**.

17. Click and drag the edge of the clip and observe its behavior.

18. Move the cursor to the upper edge of the clip to have it change to the **Fade Tool**:

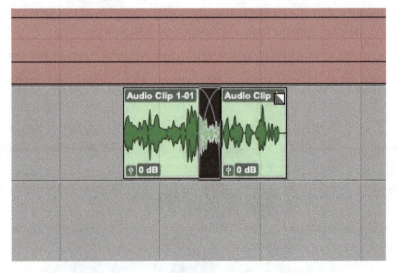

Figure 3.4: The Fade Tool

19. Click and drag from the edge of the clip toward the center of the clip to generate a fade.

20. Move the cursor to different areas of the fade and observe similar behavior to the **Crossfade Tool** previously explored.

How it works...

The **Smart Tool** changes its function and behavior according to two variables: if the cursor is in the middle or edge of a clip and if the cursor is on the upper half or bottom half of a clip. The behavior of the **Smart Tool** also changes slightly if the edge of a clip is directly next to another clip. Knowing this, you can learn all the different scenarios:

- Middle of a clip:

    - Upper half—**Select Tool**

    - Bottom half—**Grabber Tool**

- Edge of a clip (no adjacent clip):

    - Upper half—**Fade Tool**

    - Lower half—**Trimmer Tool**

- Edge of a clip (adjacent to another clip):

    - Upper third—**Fade Tool**

    - Middle third—**Trimmer Tool**

    - Bottom third—**Crossfade Tool**

The **Smart Tool** also has distinct functions according to its position within a fade:

- Upper edge—**Trimmer Tool**
- Middle—**Fade Tool**
- Bottom middle—**Grabber Tool**

Since how a fade behaves after an edit changes according to whether you are adjusting the inside or outside edge, the best way to see the results of a change is to play with a fade as in the steps listed previously and observe the results.

There's more...

Both the **Trimmer Tool** and the **Grabber Tool** have multiple sub-tools, such as the **Trimmer Tool's Time Compression/Expansion Tool**. These can be selected by clicking and holding on the tools, or by pressing their corresponding hotkey multiple times to cycle through them. When engaged, the **Smart Tool** will respect these sub-tool selections, except for the **Grabber Tool's Object Tool**, which will revert to the normal **Grabber Tool**. Keep this in mind while editing with the **Smart Tool**, as you might accidentally stretch or loop a clip when you meant to trim it.

Editing with the keyboard

As with most professional software applications, the real power of using it becomes apparent when you learn the various keyboard shortcuts to perform actions. Unlike most other software, Pro Tools binds many actions to single keys as opposed to mostly using combinations with modifier keys. Learning these commands and how keyboard focus functions is the one area that will impact your proficiency in using Pro Tools more than anything. There are many, many keyboard shortcuts and hotkeys, but for this recipe, we will focus on the keys that impact editing specifically.

Getting ready

For this recipe, you will need a Pro Tools session with an audio clip placed into a track. The commands in this recipe require **Keyboard Focus** to be active in the **Edit** window. This can be done by clicking the small **az** icon on the top right of the **Edit** window. You can also set the keyboard focus to the **Edit** window with *command + Option + 1* (*Ctrl + Alt + 1* on Windows). If **Keyboard Focus** is disabled, you can perform the actions with a hotkey by using *Control* (*Start* on Windows) as a modifier. For example, to add an edit with **Keyboard Focus** enabled on the **Edit** window, you simply press *B* on your keyboard. To add an edit while **Keyboard Focus** is active on another panel, you would press *Control + B* (*Start + B* on Windows).

This workflow will also work best if **Tab to Transients** is disabled. You can find this on the bottom left of the tool selector section of the **Tool Bar**, or with *command + Option + Tab* (*Ctrl + Alt + Tab* on Windows). **Insertion Follows Playback** should also be active. This is the button below the **Select Tool** and can be activated/deactivated with *N*:

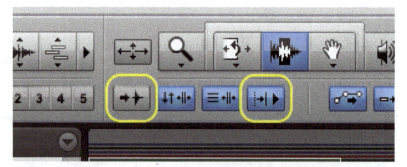

Figure 3.5: Tab to Transients and Insertion Follows Playback

Some of the hotkeys require the function or *F* keys on your keyboard. If you are using a Mac, depending on the model these keys may not be present (with the control strip on certain MacBooks, for example) or might be reserved for system commands, such as for brightness and volume. To make these keys available, go to **System Preferences**, and in the **Keyboard** tab, you should find an option for **Use F1, F2, etc. keys as standard function keys,** or you will need to add Pro Tools to the list of applications that use **Function Keys** on the **Shortcuts** tab.

How to do it...

In this recipe, we'll play around with some of the hotkeys, navigate around the **Edit** window, and perform some edits on a clip:

1. Activate the **Select Tool** with *F7*.

2. Click anywhere toward the beginning of a clip.

3. Press *A* to move the start of the clip to the selection point.

4. Press *space* to start playback and press it again to stop it further into the clip.

5. Press *D* to add a fade from the start of the clip to the selection point.

6. Move the selection point further into the clip once more by starting and stopping playback with *space*, or simply click on another spot with the **Select Tool**.

7. Press *G* to add a fade from the end of the clip to the selection point.

8. Move the selection point to somewhere inside the fade.

9. Press *S* to move the fade's end point (and the clip) to the selection point.

10. Place the selection cursor somewhere in the middle of the clip.

11. Press *B* to insert an edit at the selection point.

12. Zoom out and in with *R* and *T* respectively.

13. Place the selection cursor somewhere later in the track.

14. Press *F8* to activate the **Grabber Tool**.

15. Hold *Control* (*Start* on Windows) and click on the second half of the clip to move its start point to the selection position.

16. Move the selection cursor to another location in the track.

17. Hold *command* + *Control* (*Ctrl* + *Start*) and click on the other half of the clip to move its start point to the selection location.

18. Press *L* to move the selection cursor to the previous edit (left).

19. Press ' to move the selection cursor to the next edit (right).

20. Press *P* to move the selection cursor up a track.

21. Press ; to move the selection cursor down a track.

22. Press *F7* and *F8* at the same time to activate the **Smart Tool**.

23. Position the mouse cursor over the bottom middle of a clip to have it change to the **Grabber Tool**, and click on the clip to select it.

24. Press , to nudge the track left or earlier in the timeline.

25. Press . to nudge the track right or later in the timeline.

26. Both *M* and / will nudge the track left or right at a larger interval.

How it works...

There are a lot of commands and hotkeys covered in this recipe, so I like to break them down into groupings to make them easier to remember.

Activating tools

The *F6*, *F7*, and *F8* function keys represent the most common tools used in Pro Tools—the **Trimmer Tool**, **Selector Tool**, and **Grabber Tool** respectively. Pressing one of these keys when the tool is already active will progressively cycle through the sub-tools available for each set of tools (for example, the **Trimmer Tool** becomes the **Time Compression/Expansion Tool** when *F6* is pressed twice). *F5* and *F9* also represent the **Zoomer Tool** and **Audio Scrubber Tool** respectively, which makes sense as they are the tools directly left and right of the main editing tools. Pressing any combination of *F6*, *F7*, and *F8* will activate the **Smart Tool**.

Changing clip start and end points

The letters on the keyboard from *A* to *G* represent the main set of tools that adjust the start and end points of a clip. Using *A* will adjust the start point, and *S* will adjust the end point. These work for the start and end of fades as well. The *D* and *G* keys create fades from the start or end points of a clip. They can be used to adjust a fade end point if it's at the start of the clip and the start point if it's at the end of the clip. The *F* key can also generate fades, but a selection range needs to be highlighted first (more on this in the next recipe). Finally, the *B* key adds an edit, which is also known as separating the clips. Using these hotkeys in combination with the spacebar to start and stop playback is one way to speed up your workflow.

Nudging clips

The keys from *M* to / are what I see as the group of nudge keys, with the inner keys (easier to remember as they are also < and >) being a small nudge and the outer keys being a larger nudge. The inner keys can be adjusted by **Nudge Value** in the **Tool Bar**. You can select the time base that makes the most sense for your project. For example, when working with music, you might want to use **Bars|Beats**, and for motion pictures, you most likely want to use **Timecode**. You can also select from the dropdown next to **Nudge Value** the most common nudge amounts used in that time base, such as frames for **Timecode** and notes for **Bars|Beats**. If you require a larger or very specific nudge amount, you can simply type it into the **Nudge** window:

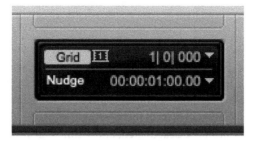

Figure 3.6: The Nudge window

When using the outer keys, the actual amount will be the next value up in the **Nudge Value** dropdown. So, if you are in **Timecode** and you have 5 frames selected for your nudge amount, m and / will be 1 second. The highest value in this drop-down menu will dictate how high this larger nudge amount will be, though, so if you have your nudge set to 1 second or higher, the outer nudge keys will still be 1 second.

Nudging can move selected clips back and forth, but it can also move the selection insertion of that value as well. This is extremely helpful for navigating the timeline or when wanting to make incremental movements to the selection location before you commit an edit. For example, if you used the method described in the previous *Changing clip start and end points* section and were listening to playback to decide where to have an edit placed, but you overshot the stop playback and needed to reverse the location by a small margin, instead of trying to find the correct spot again, you could use the nudge keys when nothing was selected and move the insertion line back and forth by a small amount.

Navigating tracks and clips

I like to look at the group of keys used for navigating the timeline as arrow keys on your keyboard. The L and 'keys are for left and right, while P and ; are for up and down. The up and down commands will typically always behave as you expect. If you are in a track lower on the timeline and press P for up, the selection point will move up a track. Left and right are a little less obvious the first time you use them, as they don't move an absolute value like the **Nudge** tools, but instead will move to the next edit point. This could be the start or end of a clip or the start/end of a fade. So, if there is a single clip with a fade at both the start and end of the clip, and the selection point is outside of the clip, it will take four presses of one of these keys to move the selection cursor all the way across the clip. If there is an edit point or separation in the clip, this will also add to the number of key presses needed to move across it.

Using these keys can also navigate across large parts of the timeline as it will move to the next clip regardless of how far away it is. If you are all the way at the end of a timeline in a project and there are no other clips on a track before the selection location, pressing L will move the timeline insertion all the way to the beginning of the track. This can be useful if you need to move the timeline far ahead or are looking for clips within a track as well.

There's more...

The commands listed are not exhaustive by any means, but they are the ones I have used most often during an edit, and I find them vital for general editing workflows. Certain commands can also be created by other combinations. For instance, moving a clip's start point to the selection location can be done by clicking on a clip with a modifier key held (as listed in this recipe's *How to do it...* section), but you can also use the *H* and *K* keys to perform the same task. However, this requires that the **Link Timeline and Edit Insertion** button be deactivated, and I almost always have that active for my workflows, although it can still be handy if you want to paste a clip and then immediately shift its end point to where it was pasted with the *K* key.

There are also similar commands available with more traditional modifier keys on Mac systems. For example, adding an edit or separating a clip can be performed with the *B* key, or with *command + E*. Similarly, zooming in/out can be done with *R* and *T*, or you can use *command + [* and *command +]* for the same effect. It is ultimately up to you to decide which method is best suited for your workflow. I personally have an easier time remembering *command + E* with the mnemonic "E for edit," but the letter *B* could also be thought of as the "blade" key.

There are resources that list all these commands (see the *See also* section, next), but you can also look at purchasing special keyboards that have all the main hotkeys shown as special icons on the keys. Some hardware options can be expensive, but there are also silicone overlays or stickers that you can purchase that are much more affordable and help visualize those "groupings" of keys.

See also

The Pro Tools shortcuts website is a fantastic listing of all commands in Pro Tools and their keyboard bindings. It has listings for both Mac and Windows keyboard layouts and a very good search function. If you find yourself performing a task often and want to find out if a shortcut exists to make it more efficient, check it out at `http://www.protoolskeyboardshortcuts.com/`.

Generating and manipulating fades

It is impossible to overstate the usefulness of fades in audio editing. In Pro Tools, fades can be applied using volume automation, but are far more functional with the **Fade** tools. These tools create a boundary around the start, end, and overlapping areas of a clip that can be easily manipulated with editing tools and are non-destructive. As their name suggests, they can fade in a clip, fade out a clip, or cross-fade two adjacent clips.

Fades are vital for providing a smooth listening experience and are not just used for creative effects. Being able to quickly and effectively generate and manipulate fades will improve your editing workflows and deliver a better end product. I am not exaggerating when I say that almost every audio clip in all my projects begins and ends with a fade.

Getting ready

This recipe requires a Pro Tools session with multiple audio clips placed on multiple tracks. You will need to have at least two clips in separate tracks overlap each other for a short duration. **Keyboard Focus** should be active in the **Edit** window for this recipe.

How to do it...

This recipe will be going over generating fades in a few different methods. Follow along with these steps:

1. Activate the **Select Tool**.

2. Position the cursor somewhere outside a clip and click and drag to partially inside the clip to highlight a portion of it.

3. Press *F* to generate a fade.

4. Press *command + F* (*Ctrl + F* on Windows) to open the **Fade** window:

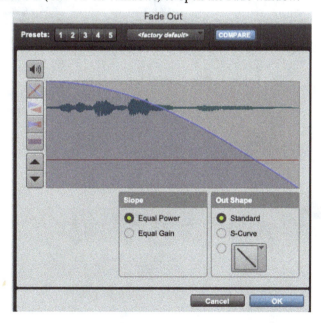

Figure 3.7: The Fade window

5. Change the **Slope** value from **Equal Gain** to **Equal Power** or vice versa.

6. Click **OK**.

7. Add some edits to another clip with *B* or *command + E*.

8. Click anywhere on a track and press *command + A* (*Ctrl + A* on Windows) to select all the audio clips in that track.

9.   Press *command + F* to bring up the **Batch Fades** window:

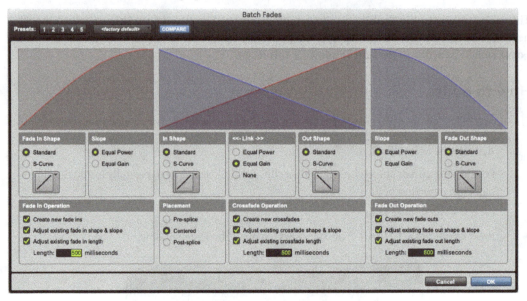

Figure 3.8: The Batch Fades window

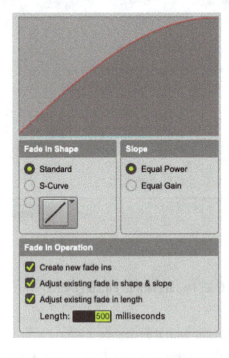

Figure 3.9: Fade In settings

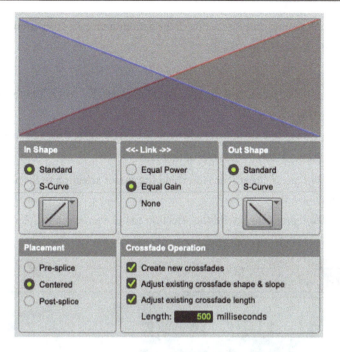

Figure 3.10: Crossfade settings

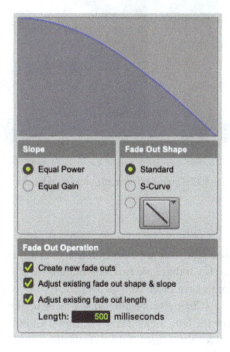

Figure 3.11: Fade Out settings

10. Leave all the default settings, but ensure that all the options under the **Operation** areas are enabled.

11. Set all the lengths to 500 milliseconds.

12. Click **OK**.

13. Locate two clips in adjacent tracks that overlap partially (or relocate two clips to be overlapped partially).

14. Click on an empty area in a track before the second overlapping clip's start point.

15. Press ' to move the selection to the start of that clip.

16. While holding *Shift*, press either *P* or *;* to extend the cursor selection to the adjacent track that has the overlapping clip.

17. While holding *Shift*, press ' to extend the selection to the overlapping clip's end.

18. Press *F* to generate a fade on both clips:

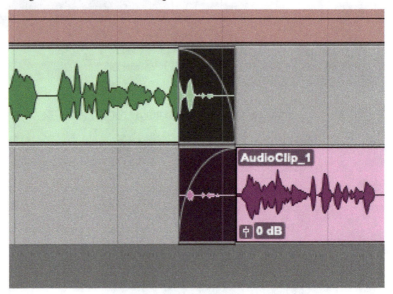

Figure 3.12: Cross-fading across multiple tracks

How it works...

Using the *F* hotkey will generate a fade over any selection that overlaps a clip's start or end point. It will not work in the middle of a clip. If there are two adjacent clips within the selection, then *F* will generate a crossfade. If you want to have more control over the type of fade being generated and its shape, then pressing *command* + *F* (*Ctrl* + *F* on Windows) will bring up the **Fade option** dialog where you can change the shape of how the fade looks. Typically, **Equal Gain** slopes are used for fade-ins and fade-outs, whereas **Equal Power** slopes are used for crossfades. The difference in sound between these **Fade** slopes is noticeable when trying to cross-fade music or other sounds that are consistent in level (such as certain ambiances or sound effects). A fade with **Equal Gain** will have a noticeable dip in the volume, whereas an **Equal Power** fade will be much more consistent. You can even see it visibly represented within the waveforms generated on a clip in Pro Tools.

When you need to generate or manipulate a large number of fades, selecting all the audio and pressing *command + F* (*Ctrl + F* on Windows) will bring up the **Batch Fade** dialog. Here, you can adjust settings for fade-ins, fade-outs, and crossfades separately. When I receive a dialogue edit for motion pictures or podcasts, I usually begin by selecting all the audio clips and performing a **Batch Fade** process. This also works with multiple tracks selected.

Finally, you then have two clips that overlap or "checkerboard;" using a combination of the navigation hotkeys, *Shift* and the fade hotkeys is a very fast and accurate way to cross-fade between tracks. Since you are using the bounds of the clips themselves as your points of navigation, this makes sure that your fades will always overlap perfectly.

There's more...

You can perform a batch fade by simply pressing *F* as opposed to opening the **Batch Fade** window. This will apply the default fade duration and slope that is set in Pro Tools' preferences. To adjust these preferences, in the menu bar, go to **Pro Tools | Preferences...** (**Edit | Preferences...** on Windows) and click on the **Editing** tab. In the top right, you can find buttons to change the **Default Fade Settings** options for **Fade In**, **Fade Out**, and **Crossfade**:

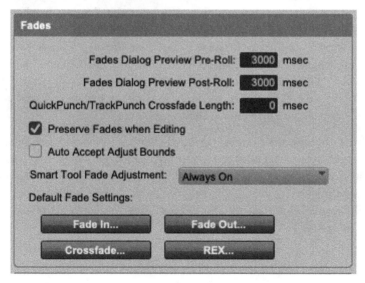

Figure 3.13: Fade preferences

Previous recipes in this chapter also define alternative ways of creating fades. Please see the *Using the Smart Tool for quick edits* and *Editing with the keyboard* recipes for other methods.

Using Edit modes correctly

I think understanding the Edit modes and how they function correctly is one of the big barriers to starting out in Pro Tools. Attempting to move and shift clips around can be frustrating if you are in the wrong mode, and the fact that the **Edit** modes are bound to the first four function keys (*F1-F4*) means you might accidentally engage in one of the modes by accident. Let's explore the different modes and what they can be used for:

Figure 3.14: Edit modes

Shuffle (F1)

Shuffle mode can be thought of as clips functioning as magnets, or like there is always a force pushing them to the left. If you take a clip in any place on any track and attempt to move it to an empty track, it will move to the beginning of the session. If there is another clip placed somewhere in that track and you drag it to somewhere after that clip, the start of the clip will jump and line up with the end of the existing clip in the track. This can be useful in some scenarios where you want to "stack" things in a sequence. For instance, if you have a long recording of an actor performing a bunch of takes of different lines, you could add edits before and after the best line, then drag it to the next track. You can continue to do this for all the lines, and you know that the track below will have all those takes lined up without any wasted space between the clips.

Another scenario where **Shuffle** mode shines is in round-table discussion podcasts. While it's possible to simply have an episode playback live to tape, removing dead air and "ums" and "uhs" is common practice and results in a tighter and easier-to-listen-to discussion. While in **Shuffle** mode, you can select all the tracks in the spot you want to remove, press *Delete* on your keyboard, and the clips will "snap" together and remove that unwanted audio without losing sync between the tracks. Combining this with **Editing Groups** can be very handy.

One other aspect of **Shuffle** mode that should be noted is the way other clips react when moving clips on the same track. If you have two or more clips on the same track, **Shuffle** mode will only let you place clips at the start or end of a clip or in between clips. The clips on the track after the clip you move will be pushed back. This is where that analogy to magnets or gravity can help a bit. For example, let's say you have three clips—we'll call them Clip_A, Clip_B, and Clip_C. If you drag Clip_C to the left slightly, an indicator will appear showing that the new position will be in between Clip_A and Clip_B. If you let go, the new order will be Clip_A, Clip_C, Clip_B, and there will be no overlapping or cut-off audio. The duration of the three clips side by side will not be affected. This mode and method can be very useful in situations where you want to reorder clips but not affect the overall timing, and combined with other tools can be very useful in music manipulation.

Slip (F2)

For those working in linear media or non-musical projects, this will most likely be the mode you end up using 99% of the time. **Slip** mode works like most conventional non-linear editors and **digital audio workstations** (**DAWs**). Clips can move freely across the timeline and into other tracks with no restrictions as to how they are placed. This can cause challenges when relocating clips across tracks if you're not careful—see the *Copying and pasting clips* recipe for more information.

There is also one other toggle to keep in mind when using **Slip** mode more than the others—**Layered Editing**. This can be enabled and disabled in the **Tool Bar** under the **Pencil Tool**, or you can go to the menu bar and select **Options | Layered Editing**. When **Layered Editing** is disabled, any time a clip overlaps another, whether it is by moving the clip or using the **Trimmer Tool** to extend one clip over the other, that area that is overlapped will be deleted. With **Layered Editing** enabled, portions of a clip's audio will not be deleted unless it is fully eclipsed by another clip. I find the latter method more functional for my editing, but there may be scenarios where you prefer to have **Layered Editing** disabled.

Spot (F3)

Spot mode is useful when you know exactly where a clip is supposed to be placed within the timeline. Clicking on any audio clip will bring up the **Spot Dialog** window, where you can type in where the clip should be moved to. You can select which time base is best for the method you are editing in, such as **Timecode** for a motion picture or **Bars|Beats** for music. The other feature that's beneficial for **Spot** mode is the ability to use the **Time Stamp** to return the clip to its original location within a project. This is most useful when a clip has been recorded within a session, has been relocated, and needs to be returned to its original location. While the **Spot Dialog** window is open, you'll see an area toward the bottom with **Original Time Stamp** and **User Time Stamp**. Click the arrow next to one of these to bring it up to the Start Point, which will set its start time to it:

Figure 3.15: The Spot Dialog window

Spot mode is also useful in motion pictures when a multitrack of audio recorded on set has been delivered. Typically, the picture edit and accompanying OMF/AAF files will only have a stereo mixdown of the audio captured, but recordists on set can record many numbers of tracks to isolate the sound from each microphone being used. You can use the timecode overlaid on the screen to synchronize the original tracks to the picture and pick out the best-sounding dialogue track.

Grid (F4)

Grid mode has two sub-modes that can be cycled through by pressing *F4* repeatedly. **Absolute Grid** creates sync points across the entire timeline using the value defined in the **Grid** setting in the **Tool Bar**. When moving a clip back and forth in this mode, the clips will always snap to the nearest grid line. This mode is useful when editing music, as you can set the grid to **Bars|Beats** and ensure your music is always synchronized to the rhythm/beat. There are situations where a clip may not start to snap to the grid, but you still want to preserve the timing. In these cases, **Relative Grid** is the best option. Clips will still move at the interval set by the **Grid** menu, but they will move relative to their start time as opposed to snapping to the grid.

When to use the Pencil Tool

Another powerful tool within Pro Tools, the **Pencil Tool** offers a few different ways to be used to manipulate waveforms and automation. You can use it to draw directly onto waveforms to repair them when issues arise. You can also draw different types of automation that can be used for different effects. While it may not be as commonly used as some of the other tools, it's still incredibly useful to know how and when to use it.

Getting ready

For this recipe, you'll need a Pro Tools session with an audio clip placed into a track. This workflow uses hotkeys, so ensure that **Keyboard Focus** is set to the **Edit** window (the small box with **az** in the top right).

How to do it...

In this recipe, we'll write directly onto a waveform and draw some volume automation using the **Pencil Tool**. Follow along with these steps:

1. Use the **Select Tool** and click somewhere in a clip where there are visible waveforms.

2. Zoom in to the clip by clicking with the **Zoomer Tool** (the magnifying glass next to the **Trimmer Tool**, or press *F5*) or pressing the *T* hotkey.

3. Continue to zoom in until the waveform is no longer shaded in and instead is represented by a solid line:

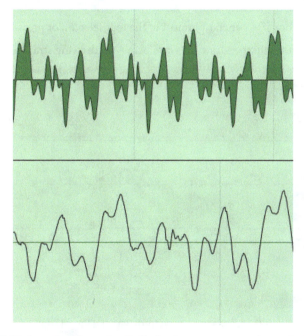

Figure 3.16: A zoomed-in audio clip (below) versus a zoomed-out clip (above)

4.   Click on the triangle on the top left of the track to open the track size menu and select **jumbo**:

Figure 3.17: The Track Height selector

5.   Select the **Pencil Tool** (the pencil rightmost of the tool selector, or press *F10*).

6.   Click and drag on the clip to redraw the waveform—try drawing straight lines or waves:

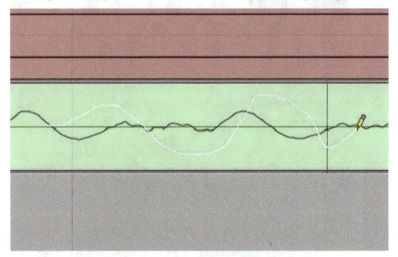

Figure 3.18: Drawing a waveform with the Pencil Tool

7.   In the track header (the left side of the tracks where the track name is listed), click on the dropdown that says **waveform** and select **volume**.

8.   Click and draw on the volume automation to draw new volume envelopes.

9.   Click and hold on the **Pencil Tool** icon and select the **Line** option, or press *F10* twice to cycle to it.

10.   Click and drag from one point to another to draw a straight line:

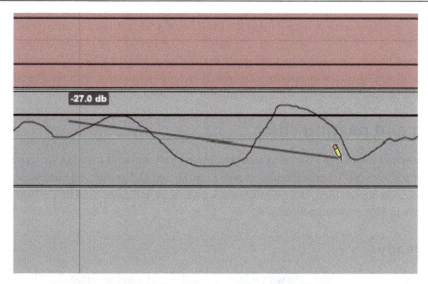

Figure 3.19: Drawing a line with the Pencil Tool

How it works...

The **Pencil Tool** will allow you to directly draw a waveform only when zoomed in far enough to no longer see the waves filled in. When zoomed in far enough, the waveforms appear as lines, which indicates you can now draw them with the **Pencil Tool**. It's very difficult to design a waveform for creative purposes with this tool. When zoomed in this far, you are looking at fractions of a second, and trying to draw anything with a specific audio result is not really possible. Instead, the **Pencil Tool** with regard to waveform manipulation should be thought of as a repair tool. Often, audio can have noticeable momentary faults—pops, clicks, or loud thuds that are incidental and don't appear regularly. When these occur, instead of using a plugin or another tool to remove the offending sound, you can simply draw over it.

This method works best when the issue is very short in duration. You can typically see the issue visually, especially with a pop or click as there is usually a noticeably sharp change in direction in the waveform as opposed to a smooth wave. Make sure to connect the start and stop of the waveform you draw to the existing wave, or you will end up introducing another sharp incline, causing yet another pop. It's not always possible to solve an issue with the **Pencil Tool**, but for those moments when it happens, it can feel almost magical.

The **Pencil Tool** is also an effective way to edit automation in a natural manner as opposed to only being restricted to using nodes. If there is a clip that requires a smooth movement of volume, for example, you can draw the volume automation with the **Pencil Tool** to have it move more naturally with a wave-like motion. You can also use the **Pencil Tool** in **Line** mode to easily draw swells from one point to another in one action as opposed to needing many automation nodes.

Finally, you can get some creative output using the **Pencil Tool** with automation. Since you can draw extreme automation paths with relative ease, you could use the **Pencil Tool** to have volume rise and drop quickly and rapidly for a tremolo-style effect, or even draw pan data, having something move left and right quickly. This type of automation is much easier to implement with the **Pencil Tool** as opposed to adding manual nodes or writing automation with other methods.

Copying and pasting clips

It might seem trivial to go over what most would think is a basic concept for a DAW—copying and pasting clips—but with Pro Tools, something as simple as copying and pasting can yield very different results depending on how it's used. Learning different ways to copy and paste clip data can help improve your workflows and keep data synchronized.

Getting ready

This recipe requires a Pro Tools session with at least two audio tracks and an audio clip. This workflow uses hotkeys, so ensure that **Keyboard Focus** is set to the **Edit** window (the small box with **az** in the top right).

How to do it...

We'll copy and paste a few different ways, including different ways of reproducing the effect of a copy-and-paste command with different tools. Follow along with these steps:

1. Use the **Grabber Tool** and click on a clip to select it.
2. Press *C* to copy the clip.
3. Press ' to move the selector cursor to the end of the clip.
4. Press *V* to paste the clip.
5. Press *L* twice to move the selection cursor to the beginning of the original clip.
6. Press *P* or ; to move the cursor up or down to the next track.
7. Press *V* to paste the clip into the track.
8. Activate the **Select Tool**.
9. Click and drag over a small portion of a clip.
10. Press *C* to copy the portion of the clip.
11. Click on another area of the track.
12. Press *V* to paste the portion of the clip.
13. Activate the **Grabber Tool**.
14. Click on one of the clips to select it.

15. Hold *Option* (*Alt* on Windows) and move the clip to another location to duplicate it.

16. Switch to the **Loop Trim Tool** (one of the **Trimmer Tool**'s sub-tools).

17. Position the cursor over the end of a clip.

18. Click and drag the clip to have it loop:

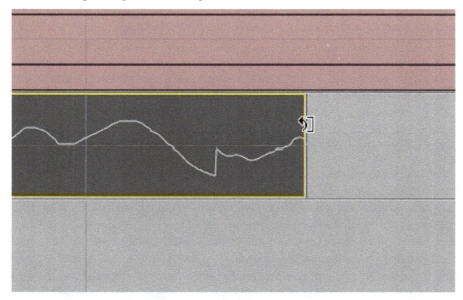

Figure 3.20: Looping a clip with the Loop Trim Tool

19. Activate the **Grabber Tool** and select another clip.

20. In the Menu Bar, go to **Clip | Loop…** to bring up the **Clip Looping** window.

21. Enter the number 10 in the **Number of loops** field.

22. Click **OK**.

How it works…

You can copy and paste whole clips using the **Grabber Tool**, or just portions of clips with the **Select Tool**. If you know where a clip is to be placed, then you can place the selection cursor in that location after you've copied a clip and paste it there, but often you will want to paste clips within certain constraints. Using the cursor navigation tools, it's possible to copy and paste clips very quickly and accurately. This is especially helpful when moving clips across tracks. While you can use *Control* (*Start* on Windows) to lock a clip's start time as you move across tracks, using the **Copy** command and *p* or *;* to move the cursor across clips and paste them where you want to is much faster and less prone to mistakes.

In situations where you want to duplicate a clip to a specific location with the mouse/trackpad, it's often easier to simply use *Option* + click (*Alt* + click on Windows) to duplicate the track. And in situations where you want the clip to repeat for a specified duration, then the **Loop** tools are your best solution.

There's more...

Copying and pasting audio data isn't just for creative situations, such as repeating musical phrases or sound effects. Copying and pasting small portions of audio can often be used in audio-repair scenarios. If there is a momentary issue with a sound, deleting it is unwise as there will be a noticeable moment of silence that can be jarring. Instead, copying a small portion of a similarly sounding audio clip and pasting it over the offending area will result in a smoother and less noticeable change. This can be a relatively long section, such as removing entire notes or words from a performance, or it can be as short as a few milliseconds to mask a pop or click. Depending on the nature of the sound, you may need to add very short fades to it to make it less noticeable.

Copying and pasting clips with minimal effort and high accuracy is also important when importing session data. While you can import audio clips directly into tracks, often you will want to be more selective both in which clips you want to import and where you want them to go. Using the *Shift* key, you can extend the range of selection for both copying clips and pasting them, even across tracks! You can copy several tracks of clips at a time and paste them elsewhere in your session. If the number of tracks you are pasting to is less than what you copied from, it will paste only the first of those tracks. For example, if you had six tracks of clips and pasted them into three destination tracks, only the first three tracks of clips will get pasted, and the last three will be ignored. On the other hand, if your selection for pasting is larger than what you copied, it will only copy what you had selected and leave the other selected tracks untouched.

Finally, keep in mind that the same behavior in placing clips exists when copying and pasting clips from tracks that don't have the same channel count. For example, you cannot paste a mono clip into a stereo track. You can, however, paste two mono clips into a stereo track. If you copy a stereo clip and try to paste it into a mono track, only the first track (the left channel) will be pasted.

Printing effects on clips using AudioSuite

AudioSuite in Pro Tools allows you to add effects and other utilities to audio clips within a timeline. These can be the stock plugins that come with Pro Tools or third-party ones added after the fact. While creating effects for sounds is typically saved for the mix phase after the clips are edited, some effects can only be done by printing them with AudioSuite. All changes made with AudioSuite are destructive in nature. In many situations, it's a good idea to preserve the original audio before any effects are applied in case you want to make changes later. Often, choices made to a sound are fine in isolation, but in the context of a mix are not the best choices and need rethinking and redoing. For this reason, we'll be going through the process of duplicating and muting the original audio before printing an AudioSuite effect.

Getting ready

For this recipe, you'll need a Pro Tools session with two audio tracks and one audio clip placed into it.

How to do it...

This workflow involves duplicating a clip into a separate track below it, muting the duplicate, and then applying an effect to the original clip (in this case, reverse). Follow along with these steps:

1. Activate the **Grabber Tool** and click on a clip to select it.

2. Duplicate the clip to the track below it by pressing *C* then *;* and then *V*.

3. Press *command + M* (*Ctrl + M* on Windows) to mute the clip.

4. Use the **Grabber Tool** to select the original clip.

5. In the Menu Bar, go to **AudioSuite | Other | Reverse**.

6. In the **AudioSuite** window that appears, click **Render**.

7. Close the **AudioSuite** window:

Figure 3.21: A clip with a printed AudioSuite effect next to its muted original clip

How it works...

While this is a relatively simple workflow, it can save a lot of headaches if you decide to make changes later. It's true that you can replace the original audio from the Clip List, but often a fair amount of editing and timing decisions go into a clip before deciding to use an AudioSuite effect. Some effects, such as **Reverse,** can only be performed in AudioSuite, so there's no choice but to use it in this fashion, which is destructive in nature.

There are a number of effects within AudioSuite that are also available as insert plugins too (plugins placed on a track to apply an effect in real time), but sometimes you only want to make changes to a single clip, so placing an effect on the whole track doesn't make sense. It's also possible that an inserted effect can be memory and processor intensive. Printed AudioSuite effects are rendered and therefore do not require computer or hardware resources while playback is operating.

There's more...

All AudioSuite effects have a few options to consider. Below the plugin selector on the top left of the **AudioSuite** window are the **create continuous file** and **create individual files** options. These also correspond to the next option—whether files are printed **clip by clip** or **entire selection**. If you have an area of your track selected that goes over multiple clips, these dropdowns will affect how the clips appear after they're rendered. Creating individual files will retain the timing and start/stop times of the original clips. Rendering the entire selection will process everything between the clips, including silence between clips if there is any. Both methods are valid, although if you are applying an effect that has some sort of decay or tail after the clip sound has ended—such as reverb, for example—then the entire selection is less likely to cut off those tails.

Next to the **Render** button is also a number you can type a value into. This value represents the handles that will be rendered before and after the clip in seconds. Handles give you some wiggle room for changes in case the head or the tail of a clip exceeds the original bounds you selected. They also make crossfading clips less likely to have invalid bounds since additional audio data will be available. Clip timing is not extended or changed when handles are added. The AudioSuite effect is printed to the length specified both before and after the clip, then the clip is trimmed back to its original size. It's always a good idea to have some extra audio to work with, so adding some handles is advised.

4

The Mechanics of Mixing

The editing stage of a project is when you decide *what* the audience is going to hear. The mixing stage is where you decide *how* they are going to hear it. The goals of a mix are not only to set the levels of each track but to also sculpt the sound and add character to what's being heard. Sometimes, a recording that was less than ideal can be saved with proper mixing techniques, and even the best-recorded sounds will suffer if not mixed well. This chapter isn't focused so much on providing you with workflows to speed up a mix—this stage of production should take time and focus. Instead, we'll explore all the different tools available to you in terms of mix mechanics. You'll learn different ways to manipulate audio levels, work with automation and **Preview** mode to audition different ideas, and how to route audio within a mix context. The next chapter will go over specific tools for shaping the sound and adding character.

In this chapter, we'll explore the following topics:

- Making coarse-level adjustments with **Clip Gain**
- Drawing volume automation to adjust levels
- Using faders and automation
- Using inserts to add effects to tracks
- Automating plugin effects
- Previewing automation for non-destructive auditioning
- When and how to use Aux tracks
- Splitting audio with sends
- Routing signal paths for a mix

Technical requirements

Most of the recipes in this chapter are designed to work with Pro Tools Ultimate, although some of the tools' features will still work with lower versions.

Making coarse-level adjustments with Clip Gain

Clip Gain is a relatively new feature to Pro Tools, although it's been present in the software for more than 10 years now. While adjusting volume is an important step in the mixing process, there are many situations where you want to adjust a clip's loudness before it goes through any other part of the signal chain. This is where **Clip Gain** comes in. We're going to use **Clip Gain** to make some coarse adjustments to a clip's loudness and look at how to manipulate it in different ways.

Getting ready

For this recipe, you'll need a Pro Tools session with an audio clip on a single track. **Clip Gain Info** also needs to be enabled. You can do this by going to the menu bar and checking whether **View | Clip | Clip Gain Info** is checked. With this active, at the start of every clip, you should see a small symbol that looks like a fader along with **0 dB** in the bottom left. Some commands in this recipe require **Keyboard Focus** to be active in the **Edit** window. This can be done by clicking the small **az** icon on the top right of the **Edit** window or pressing *command + Option + 1* (*Ctrl + Alt + 1* on Windows).

How to do it...

In this recipe, we'll play around with some clips and their **Clip Gain** settings to see how it behaves. We'll also add some **Clip Gain** automation and change the gain over time. To do this, follow these steps:

1. With any tool selected, float the cursor over the **Clip Gain** fader at the bottom left of the start of a clip.

2. Click and drag the **Clip Gain** fader and decrease it to -5 dB—you can hold *command* (*Ctrl* on Windows) to get finer control if needed:

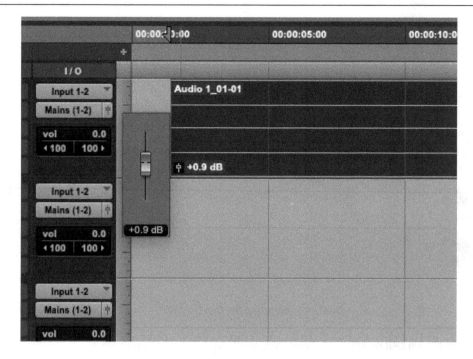

Figure 4.1: Clip Gain fader

3. Use **Select Tool** and click in the middle of the clip.

4. Press *B* or *command + E* (*Ctrl + E* on Windows) to add an edit.

5. On the new clip, click and drag its **Clip Gain** fader until it's at +5 dB.

6. Use **Select Tool** to click and drag over a section of the first clip.

7. In the menu bar, go to **Edit | Copy Special | Clip Gain**.

8. Press *command + A* (*Ctrl + A* on Windows) to select the entire track.

9. Press *V* or *command + V* (*Ctrl + V* on Windows) to paste the clip gain to all the clips on the track.

10. Go to the menu bar and enable **View | Clip | Clip Gain Line**.

11. Select the **Grabber Tool** and click on the **Clip Gain** line to add an automation point.

12. Click on two other areas to add two more points.

13. Click and drag the middle point up to increase the **Clip Gain** line at that point:

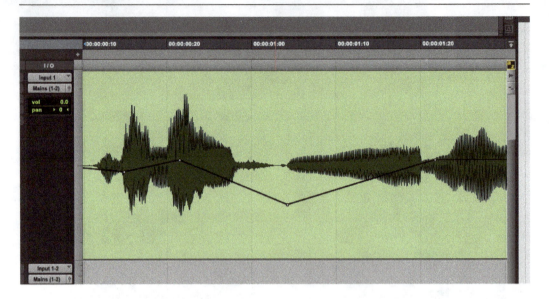

Figure 4.2: Adjusting the Clip Gain line

14. Hold *Option* (*Alt* on Windows) and click on the middle point to remove it.

How it works...

While it might seem on the surface that adjusting **Clip Gain** is similar to volume adjustments, there are a few important details that distinguish it. The most notable difference is how it affects the actual waveforms. Clip Gain adjustments affect the waveforms on the clip itself; you can see them increase and decrease as you adjust them. This can be used to make broad adjustments to *match* clips relative to one another before any other processing or volume is applied. This can also be used to significantly increase the loudness of a clip far more than volume automation (+36 dB as opposed to +12 dB, but **AudioSuite Gain** can still outperform both at +96 dB). This becomes most useful when using 32-bit (float) bit depth, as it can bring up details from very quiet recordings with much better signal-to-noise ratios.

Being able to copy and paste clip gain is very useful. You may be presented with a situation where an audio clip has been heavily edited with many cuts. Since Clip Gain is adjusted per clip, making a change on the first clip will not affect the others. Instead, using the **Copy Special | Clip Gain** command and pasting it to the target clips will make fast work of the changes you need and keep things consistent.

Finally, if you have specific moments in a clip that require a change in volume, using **Clip Gain Line** is the perfect solution to fix something at the clip level. This could be a moment that is too loud or too quiet, or it could be an audio error that you wish to drastically reduce for a short period of time. Using editing techniques can work, but sometimes **Clip Gain** will yield the best results. Don't forget about this tool in your arsenal.

There's more...

The other aspect of **Clip Gain** that is important to consider is that it will apply loudness to the audio clips before any other signal processing. If you are working with dynamic processors or other plugins that are affected by the clip's loudness, adjusting the **Clip Gain** settings will have a dramatic effect on how those plugins react.

Drawing volume automation to adjust levels

After coarse adjustments have been applied with **Clip Gain**, you can start playing with volume automation to adjust the loudness of the audio within a track over time. Volume adjustments happen after all other processing, so the effect it has is different from **Clip Gain**. Depending on your workflow and the needs of the project, it's sometimes best to simply use the mouse to add and draw automation points. We'll tackle how to do this, along with some other tools and tricks for manipulating automation after it's written.

Getting ready

For this recipe, you will need a Pro Tools session with an audio file placed into a track. Some commands in this recipe require **Keyboard Focus** to be active in the **Edit** window. This can be done by clicking the small **az** icon at the top right of the **Edit** window or pressing *command + Option + 1* (*Ctrl + Alt + 1* on Windows).

How to do it...

We're going to add some automation data to make changes to volume over time, and then play with the automation points in a few different ways. Follow along with these steps:

1. In the track header, click the button on the bottom left that looks like two circles with a line to open an **automation lane**:

Figure 4.3: The automation lane show/hide button

2. Activate **Grabber Tool** and click on the automation line at four different locations across the track.

3. Click and drag the last point down and observe how the automation line behaves.

4. Click and drag the other points and observe their behavior as well.

5. Hold *command* (*Ctrl* on Windows) and move the points up and down by clicking and dragging, and observe how they move with much finer control:

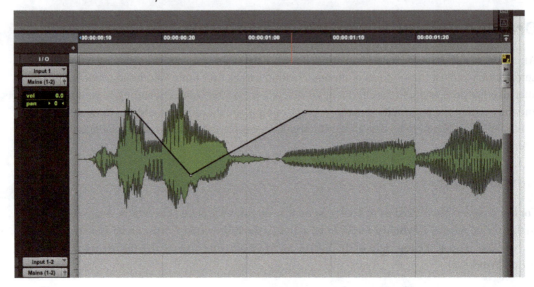

Figure 4.4: Adding automation points

6. Activate **Select Tool** and highlight the two inner automation points.

7. Press , a few times to shift the points to the left.

8. Press . a few times to shift them back to the right

9. Select **Trimmer Tool**.

10. Position the mouse cursor over the selected area and click and drag down to see how the automation points behave:

Figure 4.5: Adjusting automation points with Trimmer Tool

11. Activate **Grabber Tool**.

12. Hold *Option* (*Alt* on Windows) and observe how the icon changes to show a minus symbol.

13. While holding *Option* (*Alt* on Windows), click on the newly created automation points to remove them.

14. Activate **Select Tool**.

15. Click and drag over all the automation points to highlight them.

16. Press *C* or *command* + *C* (*Ctrl* + *C* on Windows) to copy the automation data to the clipboard.

17. Click on another location further into the track.

18. Press *V* or *command* + *V* (*Ctrl* + *V* on Windows) to copy the automation data to the clipboard:

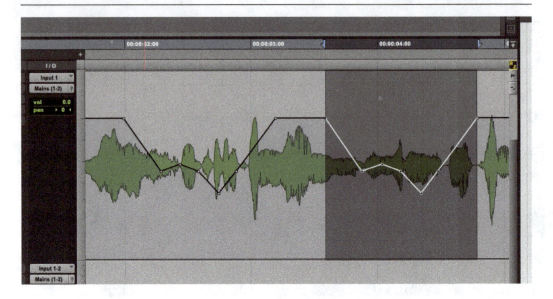

Figure 4.6: Copying and pasting automation

How it works...

Volume automation works similarly to **Clip Gain**, in that it will affect the volume of a track over time according to the points you place. Where it differs is that these volume adjustments are applied after any plugins or effects. You can easily add and remove automation points using **Grabber Tool**, and both nudging and copying and pasting automation points work as you might expect them to.

It's when you try to adjust the volume of several automation points where things can behave unexpectedly. **Trimmer Tool** is used to raise and lower a range of automation, and if no changes in volume are present, it will behave as you might expect, moving an entire section in one go. But if the automation points within the selection are not the same, then the adjustment will appear odd. This is because it is applying volume logarithmically, in order to try to retain the relationship in loudness between the points. It will also create new automation points on the boundaries of the selection.

There's more...

With Pro Tools Ultimate, if are trying to make changes across a large section of automation points and don't want to have the volume data's shape distorted, you should opt to work with **Volume Trim** instead. This automation data can be accessed in the same automation lane by clicking the **Volume** dropdown and selecting **Volume Trim**, or you can add another automation lane with the plus button on the left side of the lane.

Volume Trim adds a uniform increase or decrease in volume after the fact. What's neat about it is that you can still see your original volume automation data with all the points, and a blue line representing the change after the trim. Due to how it works, I highly recommend keeping the trim line flat and opting to use just **Selector Tool** and **Trimmer** as opposed to many automation points.

Pencil Tool is also a powerful option if you want to manually set automation points. Please see the *When to use the Pencil Tool* recipe in *Chapter 3, Faster Editing Techniques*, for more details.

Volume automation can also be manipulated using **Smart Tool**, which will change according to where in the track the cursor is placed, but in my experience, it's more effective to manually select the correct tool for the action you are performing.

Using faders and automation

Manually adding automation data is perfectly acceptable in many situations. However, it is often the case that relying on it exclusively does not provide enough precision or ultimately makes things sound too mechanical and not natural. The use of physical faders to manipulate sound has been around a lot longer than automation, but with Pro Tools, you do not need to have a physical fader controller to achieve similar results. We'll use Pro Tools' built-in tools to create smooth and responsive volume automation.

Getting ready

For this recipe, you will need a Pro Tools session with a single audio track. No audio clip needs to be placed, but having one will help demonstrate the volume automation effects. Your session will need to have the **I/O** column active. You can enable this by going to the menu bar and selecting **View | Edit Window Views | I/O**.

How to do it...

We'll now create some volume automation using the built-in fader with Pro Tools. Do this by following these steps:

1. Go to the menu bar and select **Window | Automation**.
2. Make sure **VOL** (volume) is enabled—it should be red, so click on it to turn it red if it is inactive:

Figure 4.7: Automation window (Pro Tools Ultimate)

3. In the track header, under the track name, there should be a dropdown with **waveform** selected (**Track View Selector**)—click this dropdown and select **volume** (this is not actually required, but helps you see the automation being written).

4. Under the volume dropdown, there is a dropdown showing **read** in green letters (**Automation Mode Selector**)—click this and change it to **touch**.

5. In the **I/O** column of the track header, click the **Fader** button next to the output selector.

6. Activate **Selector Tool** and click toward the beginning of the track.

7. Press **Play** on the transport or press the spacebar to begin playback.

8. As the track plays, click and drag the fader up and down and observe the volume automation line following your movement in red:

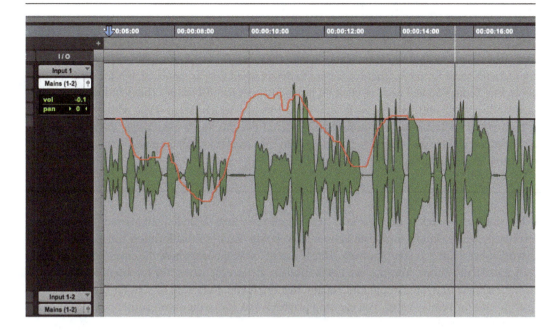

Figure 4.8: Writing volume automation with a fader

9. While the track is still playing, drag the fader up to a new position, then release the mouse click, and observe the automation returning to the previously written automation.

10. Stop playback with the spacebar and see how the automation points are now written.

11. Click on the **Automation Mode Selector** and change it to **latch**.

12. Start playback, drag the fader up or down, and release the mouse click— observe how the automation line continues to write.

13. Stop playback.

14. Change the automation mode to **write** and repeat *step 12* —observe how the automation line overwrites everything even before you click on the fader.

How it works...

For automation data to be written with a fader, you must first ensure that **Automation Write Enable** is active in the **Automation** window. Since we are exploring volume automation, we only need to make sure that the **VOL** button is active (red). Once that's confirmed, then a track can be set to one of the automation write modes. The write modes are set out next.

Touch

This mode will only write automation if you "touch" the fader. On a mix controller, faders are usually equipped with some form of capacitive feature to detect when they are touched, but for the built-in fader in Pro Tools, you will need to click on it to activate it for this mode. The defining feature of **Touch** mode is it will only write as the fader is being touched, which means if you release the mouse, it will return to the last point of automation written. I use this mode most often when using the built-in faders, and it's very useful for "punching in" to make small adjustments without affecting other automation that has already been written.

Latch

Latch is similar to **Touch** in that it won't activate until you touch the fader, but instead of returning to previously written automation data upon release, it will "latch" to that location and stay in place, overwriting everything as you play back. This is useful for situations where the entire clip needs to have a consistent volume. You can use **Latch** mode to determine where the loudness needs to be, then leave it there to write a straight line of automation. If you don't want to wait for the entire clip to play to have the result written, you can stop playback and nudge the automation points to extend to the length of the clip.

Touch/Latch

When using **Touch** mode, you may want to have other parameters stay in latch so that they adhere to the location you place them—this is where **Touch/Latch** comes into play. This is more effective on physical mixing controllers with multiple inputs.

Write

Write mode is the riskiest mode to use, as it will overwrite all automation regardless of whether you touch the fader or not. There aren't many situations where I would recommend **Write** mode, and Pro Tools will switch the mode back to **Latch** after using it once just in case.

There's more...

There is also a **Trim** option below the automation write modes. This works similarly to volume trim automation, in that it will adjust all previously written automation equally as opposed to overwriting it. This is useful when you have written a lot of automation and want to simply increase or decrease all the points at the same time.

Automation can also be set to **Read** mode once you are done with making changes to prevent things from accidentally being overwritten. When in **Read** mode with some automation data present, making a change with the fader will have a temporary effect, affecting the loudness if you make an adjustment while in playback, but no automation will be written. If no automation data is present, then the whole track's volume will be changed to whatever you set the fader to. If you want to return to this behavior, you can set automation to **off**.

Finally, you may want to have more than one fader on screen at the same time. By default, Pro Tools will replace a fader window with a new one if you click on another fader button. To keep a fader for a track active, click the red button on the top right of the fader. You will have to do this for every fader you set active. If you have multiple displays or wish to simply look at only faders, you can also activate the **Mix** window in the menu bar under **Window** | **Mix**.

The **Mix** window attempts to mimic the look of a mixing board with all the tracks laid out next to each other. The topmost track in your **Edit** window is the leftmost track in the **Mix** window. Tracks that are hidden from view in the **Edit** window are also hidden in the **Mix** window.

Using inserts to add effects to tracks

While volume automation can get you very far within the mix stage of a project, using plugins to affect the sound is how you shape and sculpt the tone as well as add dimension and character. And while it's possible to do this with AudioSuite and print changes onto clips destructively, it's a much more practical and effective use of time to use real-time plugins with inserts. With Pro Tools, any one track can have up to 10 plugins inserted, and we'll be adding some across different tracks.

Getting ready

This recipe requires a Pro Tools session with four mono audio tracks and four stereo tracks. While audio clips are not needed, having them helps demonstrate how the effect signal chain works. You will need to have **Inserts A-E** active in the track headers as well. To do this, go to the menu bar and select **View** | **Edit Window Views** | **Inserts A-E**.

How to do it...

We'll be adding plugins as inserts in a few different ways to change the way the audio sounds in real time. Follow along with these steps:

1. In the first mono track, under the **Inserts A-E** column, click the first empty insert slot.
2. Select **plug-in** | **Other** | **Trim**.
3. Click on the dot next to the **Trim** plugin in the insert slot and select **plug-in** | **EQ** | **EQ3 1-Band**.
4. Click and drag the plugin to move it to another insert slot.
5. Hold *Option* (*Alt* on Windows) and click and drag the plugin from one track to another to duplicate it
6. Hold *Option* (*Alt* on Windows) and click on any of the second insert slots and select **plug-in** | **Other** | **Trim**.

7.   Click on the first stereo track's name to highlight it.

8.   Hold *Shift* and click on the last stereo track.

9.   Hold *Option + Shift* and click on an empty insert slot.

10. Select **plug-in | multichannel plugin | Reverb | D-Verb**.

11. Hold *command* (*Ctrl* on Windows) and click on any active plugin in an insert slot to bypass it (disable it).

12. Click the dot next to any insert and select **no insert** to remove it.

13. Hold *Option* (*Alt* on Windows) and set any insert to **no insert** to remove all plugins set to inserts in that slot.

How it works...

If you've ever seen a guitar-effect pedal, you've seen a physical version of an insert. The guitar gets plugged into the pedal, and the pedal then gets plugged into an amplifier. The pedal is inserted into the signal chain, creating a step in the signal that changes how the audio sounds. In this recipe, we stuck to some of the very basic effects that come with Pro Tools, but we didn't play with their parameters (see *Chapter 5, Shaping Sounds with Plugins and Effects,* for that).

Inserts can be different, depending on the number of channels you have on a track. For instance, stereo tracks can have reverbs that are multichannel and will give a true stereo image with reflections traveling between the left and right channels. A multi-mono reverb will sound very different. Some plugins aren't available at all, depending on the number of channels you have.

Inserts follow a **top-to-down** signal path. The plugins placed at the top will process first and follow the path going down. The order in which you place the plugins can greatly affect the resulting sound. For example, **Noise Reduction** typically works best when there is a more dynamic range. If you place a compressor before a **Noise Reduction** plugin, it might not be as effective or it might introduce unwanted artifacts.

Finally, understanding all the different ways of adding inserts will allow you to more quickly set up your sessions. If you know all the dialogue or vocal tracks require a high pass filter, selecting them and holding *Option + Shift* while picking a 1-band EQ will apply that plugin to all of those highlighted. This combination is also true of many other changes to tracks in Pro Tools, including **Mute/Solo**, **Record Arm**, and changing the track view.

There's more...

Inserts can also exist as physical signal routes. Many pieces of hardware exist that provide similar effects to what's provided with Pro Tools, but these units are desired for their unique properties. You will need an audio interface that can handle inserts (with both an input and output). This is why plugins are not the only option available when you click on an empty insert slot.

There are an additional five insert slots per track available as well. You can access them in the menu bar under **View | Edit Window Views | Inserts F-J**. These still process top to down with the first column of inserts processing first.

Automating plugin effects

There are many scenarios where you may want to change the way something sounds over time. For these moments, we'll use plugin automation. Plugins can be changed over time with automation just as with volume; the difference is that instead of one fader to work with, you have a plethora of sliders and knobs, depending on which plugin you are using. Sometimes, this means only having a few parameters to consider, such as with the **Trim** plugin, for instance, or it could mean having many, many parameters, such as for an EQ. Since Pro Tools doesn't enable plugin automation by default, we're going to learn how to activate the plugin automation for each parameter and explore the different ways we can manipulate it.

Getting ready

For this recipe, you will need a Pro Tools session with a single mono audio track. No audio clips need to be present, but having one present can help demonstrate the results from plugin automation. You will need to have **Inserts A-E** active in the track headers as well. To do this, go to the menu bar and select **View | Edit Window Views | Inserts A-E**.

How to do it...

We'll now add a plugin to a track, enable automation on its parameters, and write automation in a few different ways. Follow along with these steps:

1. Go to the menu bar and select **Window | Automation**.
2. Make sure **PLUG IN** is enabled—it should be red, so click on it to turn it red if it is inactive.
3. Click on the first empty insert slot and select **plug-in | Other | Trim**.
4. In the **Trim** plugin window that appears (if it's not onscreen, click on the word **Trim** in the insert slot to bring it forward), click the button that looks like two overlapping window boxes under the **Auto** area (the **Automation dialog**).
5. In the **Plug-In Automation** window, click on the word **Gain** on the left-side list.
6. Click on the **Add >>** button.
7. Click **OK**:

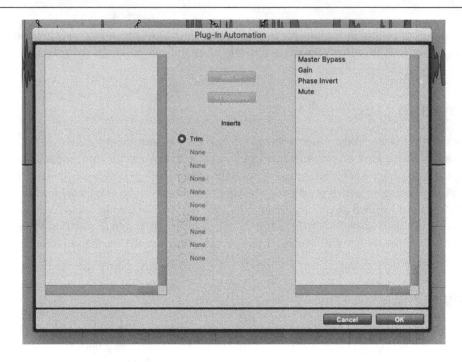

Figure 4.9: Plug-in Automation dialog window

8. Click the button in the bottom left of the track header that looks like two circles with a line drawn between them to show the automation lane.

9. Click on the **Volume** dropdown and select <<**Track Name**>> **(fx a): Trim | Gain**.

10. Activate **Grabber Tool**.

11. Add three automation points.

12. Click and drag the middle automation point up to **+6 dB**.

13. Activate **Select Tool**.

14. Click just before the first automation point you added.

15. Start playback with the transport or press the spacebar and observe how the **Gain** slider in the **Trim** plugin follows the automation points.

16. Stop playback by pressing the spacebar.

17. Right-click on the **Trim** plugin in the **Inserts** column and select **Automation Dialog…**.

18. In the **Plug-In Automation** window, click the word **Gain** on the right side.

19. Click the <<**Remove** button.

20. A warning will appear letting you know that this will delete the automation. Click **Remove** or press *Return/Enter*.

21. Click **OK**.

22. Click on the second insert slot and select **plug-in | EQ | EQ3 7-Band**.

23. Hold *Control + Option + command* (*Start + Alt + Ctrl* on Windows) and click on the **Automation Dialog** button under the **Auto** area to activate automation to all plugin parameters.

24. In the track header (outside the **Plugin** window), click the dropdown that says **read** and change it to **touch**.

25. In the automation lane, click the dropdown that says **Gain** and select **<<Track Name>> (fx b): EQ3 7-Band | Low Band Gain**.

26. Make sure the EQ plugin window is active and start playback with the spacebar.

27. While playback is active, click and drag the large red knob in the bottom left of the EQ3 plugin window up and down (it's in the area labeled **LF**—the **GAIN** knob) and observe the automation data being written to the track in the automation lane (the line turns red as it's being moved).

28. Click on somewhere to move the timeline insertion cursor to before the automation you just wrote.

29. Press the spacebar to start playback and observe the automation in action.

30. Stop playback by pressing the spacebar.

How it works...

Every knob, slider, and button within a plugin's interface (basically, anything you can change) can be represented by a number or state. Plugin automation focuses on changing that parameter over a period of time. Sometimes, you may only want to adjust a few of all the available parameters, or even just one. When that's the case, it's best to open the automation dialog from either the plugin window or by right-clicking on the plugin within the **Inserts** column and just add the parameter you wish to make adjustments to. If you're not sure which exact parameter you will be adjusting or want the flexibility of making different types of adjustments later, then you can simply add all the parameters or use the shortcut shown in the recipe to add them all in one click. The disadvantage to doing this is it will bloat the drop-down list for automation view within the track.

You can always manually make adjustments within an automation lane to a plugin parameter once automation is active on it, but if you want to use the plugin itself to make changes, you will need to make sure the automation is in a write mode. Refer to the *Using faders and automation* recipe from earlier in this chapter for more details on write modes.

There's more...

Automation data is written to the plugin, not the track. If you move or duplicate plugins to a different track, it will bring all the automation points over with it. You can use this to your advantage, but it can also be a hindrance if you're not expecting it.

You can also disable this behavior by toggling **Automation Follows Edit** under the **Options** menu (this can also be done by clicking the button directly below **Grabber Tool**), but this means moving a clip back or forth will not have the automation follow it, so proceed with caution when turning off this feature.

Previewing automation for non-destructive auditioning

Being able to write automation as you listen is vital to many mixing applications, but you will encounter situations where you need to make changes to multiple parameters at once. An EQ is a good example. While some moments might require just one frequency to be increased or decreased, more often, you will need to adjust multiple bands to create a proper curve to achieve the result you are looking for. The solution for this is **Preview Automation**. Using the preview function will allow you to make non-destructive changes to different plugins and dial in the sound you are looking for. Once you've found the settings that work best for your scenario, you can then confirm that by writing the automation to a selection. We'll be making several changes to an audio clip to demonstrate how to do that.

Getting ready

This recipe requires a Pro Tools session with a single mono track and an audio clip placed on it. The track will need both an **EQ3 7-Band** and a **Dyn3 compressor/limiter** placed on inserts with all their automation enabled and the track in automation write mode—see the *Automating plugin effects* recipe if you are not aware of how to do this. Make sure the **Inserts** and **I/O** columns are active in the track headers. You can do this by going to the menu bar and selecting **View | Edit Window Views** and selecting them.

How to do it...

We're going to audition changes to a couple of parameters in a track, and then write those changes to the automation. You can do this by following these steps:

1. Go to the menu bar and select **Window | Automation**.
2. Ensure that **PLUG IN** and **VOL** (volume) are both enabled (colored red).
3. Disable **Timeline Follows Insertion** under **Select Tool** or press *N*.
4. Press the **Preview** button.
5. In the **Inserts** column, click on **EQ3 7-Band** to activate the plugin window.
6. Activate **Select Tool** and click on the track at the beginning of the audio clip.
7. Start playback with the transport or press the spacebar.
8. Make changes to the EQ curve by clicking and dragging different knobs or the colored EQ points on the graph.

9. Press the spacebar to stop the playback.

10. Click on **Dyn3 Compressor/Limiter** in the **Inserts** column.

11. Start playback.

12. Make changes to the compressor by clicking and dragging different knobs.

13. In the **I/O** column, click and drag the **vol** button (volume) to make changes.

14. Use **Select Tool** to click and drag over a portion of the audio clip.

15. Go to the menu bar and select **Edit | Automation | Write to All Enabled** (*command + Option + / on Mac or Ctrl + Alt + / on Windows*).

How it works...

Putting automation into **Preview** mode prevents changes being made to parameters from being written to automation. This allows you to experiment and adjust a plugin parameter placed in a track and tweak it until it sounds correct. Without a preview, any changes made would record to automation over time, and you wouldn't be able to make changes to multiple parameters as they would reset once you let go (in **Auto Touch** mode, at least).

This can be useful for many different audio projects, and it is most definitely necessary for motion picture work. Since recording setups, locations, microphone types, and so many other variables change from shot to shot, being able to apply unique settings for each clip becomes vital. In other mediums, such as music production, being able to write automation from clip to clip may not be as important, but being able to preview what something sounds like non-destructively is very useful.

There's more...

There are other ways of achieving similar results with automation tools. Before **Preview Automation** was introduced into Pro Tools, using the **Suspend** button was the only way to non-destructively preview automation. Many engineers still prefer this workflow, but I personally prefer the **Preview** button since it allows other plugins and parameters to operate as you would expect them. Suspending automation turns off everything on all tracks, which can cause unexpected results.

Also, note that there are alternative ways to write automation that has been previewed. These tools are available in the **Automation** window with the different methods shown with buttons and tool tips. You can experiment with these to see how they function.

When and how to use Aux tracks

Inserts work well for adding different effects to a single track, but there are many situations where you will want to have different audio tracks be affected in the same way. This is where an Aux (auxiliary) track comes in. Audio clips cannot be placed in an Aux track; they exist solely as a location to route audio to. This allows you to combine many signals, and even duplicate signals as inputs if needed. We'll explore what this looks like practically.

Getting ready

For this recipe, you'll need a Pro Tools session with several tracks containing audio clips. Make sure the **Inserts** and **I/O** columns are active in the track headers. You can do this by going to the menu bar and selecting **View| Edit Window Views** and then selecting tracks.

How to do it...

For this recipe, we'll be taking multiple tracks and mixing them down into a single Aux track. We'll then apply some insert effects to the Aux track with plugins and hear the results. Follow along with these steps:

1. Click on the name of the first track in your session to highlight it.

2. Hold *Shift* and click on the last track.

3. Hold *Option* (*Alt* on Windows) and click on the second drop-down menu in the **I/O** column (usually labeled **1-2**) of the first track.

4. Click on **new track...**.

5. In the **New Track** window, select the following settings:

    Format: Stereo

    Type: Aux Input

    Time Base: Samples

    Name: Submix

6. Click **Create**.

7. In the newly created Submix track, click on the first empty slot in the **Inserts** column.

8. Select **multichannel plugin | Harmonic | Lo-Fi**.

9. In the **Lo-Fi** plugin window, click the drop-down menu below the **Preset** column that shows **<factory default>**.

10. Select **Factory Settings | Bass Dirty Amp**.

11. Begin playback with the transport or press the spacebar to listen to the result.

12. Adjust the different parameters in the plugin and observe the change in audio.

13. Press the spacebar to stop the playback.

How it works...

The **I/O** column in the track's header allows you to set both the track's input and output. Holding *Option* applies a change to all tracks, so highlighting all the tracks and setting one of their outputs while it's held will set all their outputs to the same setting. Setting the output to a new track then prompts you to configure what that track's settings will be. Since we set it to an Aux track, everything is set up for you when the track is created, including the Aux track's input setting.

What now happens is all the signals get mixed down to that single Aux track. Any changes applied to it, including adding insert effects, will be applied to the mixed-down signal. By using a very noticeable effect (Lo-Fi), you can hear how it is applied to everything.

This is useful not only in dramatic effects such as Lo-Fi but also with more utilitarian ones such as dynamics and EQ. The way that audio signal paths work, applying a compressor to a mixdown of many tracks will yield a different result from pplying a compressor to many tracks and then mixing them down. This also saves on the CPU if using a resource-intensive plugin, rather than having many instances of it. You still must make sure it's appropriate for the situation, though. Something such as **Noise Reduction** wouldn't typically be applied to an Aux track with signals mixed down; it's usually more effective to apply it to individual tracks.

There's more...

Using the **Mix** window (under **View** in the menu bar) is often an easier way to see all the routing at a glance. The same information and columns from the **Edit** window are available, just set out vertically.

Aux tracks can also be duplicated to try different settings out. Right-click on the Aux track and select **Duplicate...**. Type in the number of duplicates you want in the next window, then click **OK** (all the other checkboxes can remain checked). You can add as many inserts and plugins as you like and then toggle the mute button on each one (the button in the track header named **M**) to audition the difference. Sometimes, you can have multiple tracks with multiple effects playing simultaneously to get great results as well. Try experimenting to see what kinds of results you get.

Splitting audio with sends

Sending entire track audio to a send is fine if you want the sound to totally be affected by the Aux it's being routed to, but what if you want to also have an unaffected version of the audio still be present in the mix? A good example of this is when adding reverb effects to a track. While there is usually a wet/dry mix parameter that can be adjusted, to sound natural, you typically want to have a completely dry vocal mixed in with a complete wet reverb. This is where sends are used. Sends allow you to split a signal and send it to several different tracks. You can have a clean signal remain in the main mix, while also splitting it and sending that second signal to an Aux track. We'll learn how to do this in this recipe.

Getting ready

This recipe requires a Pro Tools session with a single audio track and an audio clip in the track. Make sure the **Inserts**, **Sends**, and **I/O** columns are active in the track headers. You can do this by going to the menu bar and selecting **View** | **Edit Window Views** and then selecting a track.

How to do it...

We'll take a signal from a track, send it to an Aux track, and then add a reverb plugin to the Aux track to get more natural-sounding reverb. Follow along with these steps:

1. In the track header, click on the first available **Send** slot under the **Sends** column (the button labeled **a**).

2. Select **new track…**.

3. In the **New Track** window, select the following settings:

    Format: Stereo

    Type: Aux Input

    Time Base: Samples

    Name: Verb

    Notice a new fader pops up for the send.

4. In the **Verb Aux** track, click on the first **Insert** slot and select **multichannel plug-in** | **Reverb** | **D-verb**.

5. In the **D-Verb** plugin window, click on the dropdown directly under the **Preset** column.

6. Select **Medium Room**.

7. Start playback with the transport or press the spacebar.

8. Slowly raise the **Send** fader and stop when you feel it's a good amount of reverb:

Figure 4.10: Send fader

9. Stop playback with the spacebar.

How it works...

Sends split the signal to the destination of your choice. Since we selected a new track, an Aux track was created with its input already set correctly. By default, all sends will have their output levels set to silent. You need to increase the level to hear the result. What you are hearing is a mix of the source audio from that track being sent to the main outputs and to an aux track simultaneously.

You also need to have something applied to that track to hear any changes—in this case, we added a reverb insert. We selected a **Medium Room** preset, which tries to emulate the sound of a bedroom or living room. In a real-life scenario, the microphone would most likely be up close to the performer providing that warmth and presence you'd expect to hear, but the mic would still capture the reflections of that person's voice from the walls in the room. By using this method, the listener will hear more natural-sounding reverb.

Sends work post-fader by default. If you apply any inserts to a track with a send on it, the signal being sent will have those effects applied too. Since it is post-fader, if you turn down the volume automation, the send will also be turned down.

There's more...

Sends can be used for more utility applications as well. It's a good practice to test for monoaural compatibility with a mix, regardless of the medium. Despite the prevalence of multichannel devices, there are still many times when a signal will get mixed down to mono against the wishes of the engineer who mixed it. It's therefore a good idea to make sure you are not getting any phase issues when mixed down by testing the signal in a mono setup. You can do this easily by sending your audio to a **Mono Aux** track. Simply unmute it when you want to test how things will sound folded over.

While post-fader is the default for a send, you can also click the **PRE** button on the send fader to have the send ignore the fader. Insert effects are still applied. There are some scenarios, such as when auditioning audio effects, where having sends act pre-fader is preferred.

Routing signal paths for a mix

To prepare a session for an effective mix, you'll need to consider the routes the audio's signal paths take. It is rare that all tracks will send their audio to one singular output; even in non-studio scenarios, a lot of routing takes place to offer better control over the mix, and more flexibility when making changes and adjustments. Understanding how and where to set audio paths will take the guesswork out of your mixes and allow you to make more informed decisions on how a mix is performed.

Getting ready

For this recipe, you'll need a Pro Tools session with 12 audio tracks with audio clips placed in them. Make sure the **Sends** and **I/O** columns are active in the track headers. You can do this by going to the menu bar and selecting **View | Edit Window Views** and selecting tracks.

How to do it...

We're now going to set up bussing routes using session I/O and then apply the inputs and outputs to appropriate tracks. Follow along with these steps:

1. Go to the menu bar and click on **Setup | I/O…**.
2. In the **I/O Setup** window, click on the **Bus** tab.
3. Click the **Default** button on the bottom left of the window (make sure **All Busses** is selected from the drop-down menu next to it).
4. Double-click on **Bus 1-2** and type in Submix 1.
5. Press *Return/Enter* to accept your changes.

6. Continue double-clicking and typing in names for each set of busses and name them as follows:

- `Submix 2`

- `Submix 3`

- `Verb`

- `Delay`

- `Master Mix`

7. Click on the next default bus to highlight it.

8. Hold *Shift* and click on the last default bus.

9. Click the **Delete Path** button above the **Default** button:

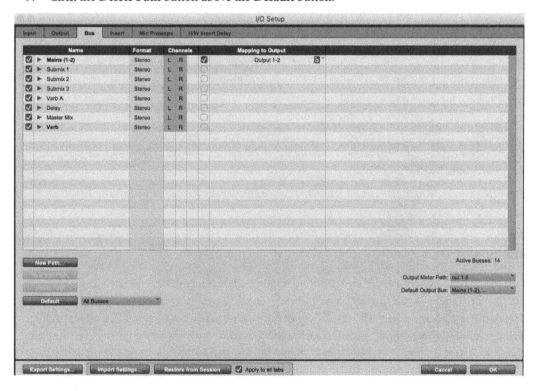

Figure 4.11: I/O Setup window

10. Click **OK**.

11. In the menu bar, go to **Track | New...**.

12. Create six stereo Aux input tracks in samples (you can leave the default name).

13. Double-click on the first Aux track to bring up the track name window.

14. Type in `Submix 1`.

15. Click the **Next** button or *command + right-click* (*Ctrl + right-click* on Windows).

16. Name the rest of the Aux tracks to match the bus names we created earlier.

17. In the **I/O** column for **Submix 1**, click the first drop-down menu (it should read **no input**).

18. Select **bus | Submix 1**.

19. Set the inputs for the remaining Aux tracks to match their track names.

20. Click on the first audio track in your session.

21. Hold *Shift* and click on the fourth audio track.

22. Hold *Option + Shift* (*Alt + Shift* on Windows) and click on the second dropdown in the **I/O** column.

23. Select **bus | Submix 1** (make sure to select the main stereo path, not the mono sub-paths listed beneath it).

24. Repeat *steps 20-23* for the next set of four audio tracks, setting their output to **Submix 2**, then repeat this for the last set of audio tracks to go to **Submix 3**.

25. Click on the first audio track.

26. Hold *Shift* and click on the last audio track.

27. Hold *Option + Shift* (*Alt + Shift* on Windows) and click on the first **Send** slot (**a**).

28. Select **bus | Verb**.

29. Hold *Option + Shift* (*Alt + Shift* on Windows) and click on the second **Send** slot (**b**).

30. Select **bus | Delay**.

How it works...

Busses are virtual signal paths. They do not go to any track by default, so you have to set a track's input to accept them. When you set an output or send it to a new or existing track, it creates a bus to match the track name unless a bus is already being used on that track's input. Manually setting up busses prior to assigning outputs is beneficial for being exact in terms of how audio will be routed as opposed to relying on Pro Tools' bus-naming methods. I've often seen sessions where multiple busses were named haphazardly, making it harder to assign signals correctly.

Busses can also allow for more advanced signal splitting and routing. As we demonstrated, you can have multiple tracks send their signal to the same bus. Similarly, a bus can be assigned to multiple inputs. This can be great for layering effects on different tracks.

Finally, setting up sessions to have submixes can provide easier auditioning. For example, if I were mixing a song, I might have submixes labeled for drums, guitar, vocals, and so on. If I wanted to focus on just the backing tracks, I could mute the vocals in one click if they were routed as such. Muting one submix is much easier than having to mute/unmute many tracks that are routed to it.

There's more...

When working with submixes and Aux tracks, you'll probably want to have solo safe enabled on the Aux tracks. To do this, hold *command* (*Ctrl* on Windows) and click on the **Solo** button (the **S** beneath the track name). You will see the **Solo** button become grayed out. This will make the track ignore the **Solo** buttons from other tracks and always remain active. Without this enabled, if you solo a track that's routed to the aux, it won't play because the aux track will be muted.

Just as with setting up routes, you can hold *option + Shift* (*Alt + Shift* on Windows) to apply **Solo** safely to all selected tracks. This means actually holding *command + Option + Shift* (*Ctrl + Alt + Shift* on Windows) while clicking on a track's **Solo** button.

It's also important to keep in mind that when creating a new session, the default option for session I/O is to simply use the last session settings. If you are consistently working in the same medium or area of sound and need the same routing routinely, there's nothing wrong with simply using the last session's I/O settings. If you accepted that option without understanding the consequences, you could wind up with a very confusing I/O setup.

5

Shaping Sounds with Plugins and Effects

Volume is an integral part of the mixing process, but the most creative and challenging parts of any mix are deciding which plugins and effects to use and how to apply them. You can use plugins to repair audio that has technical problems, and you can also shape the way they sound and interact with each other. Used to good effect, you can invoke powerful reactions from the listener and provide information to the audience that wasn't there in the audio on its own. In this chapter, we'll look at the fundamental tools used to mold and sculpt sounds and how to combine them for specific use cases. You'll learn when and why to apply things such as **Equalization** (**EQ**), and how to best use it for different situations. All the examples shown will use stock Pro Tools plugins, but we'll also touch on some of the other popular plugin offerings by third-party vendors.

In this chapter, you will learn about the following key topics:

- Cleaning up sounds with Noise Reduction and Expanders
- Using EQ to shape the tone of a sound
- How to clean up hum and crackle with Notch Filters
- Using Reverb to create space
- Making things "strange and scary" with reverse reverb
- Using Compression to control dynamics
- Advanced sound shaping with Multiband Compression
- Making sounds louder without killing dynamics with Parallel Compression
- Printing Pitch Shift effects with AudioSuite

Technical requirements

This chapter requires you to use at least the Pro Tools Artist version for most recipes, with some recipes benefiting from the extra plugins offered in Pro Tools Studio.

Cleaning up sounds with Noise Reduction and Expanders

It is almost impossible to get a perfect-sounding recording right out of the gate. Noise in its many forms tends to find its way into audio recordings even under the best circumstances. With that in mind, being able to clean up unwanted noises present in audio clips is one of the fundamental skills needed when mixing.

Getting ready

For this session, you will need a Pro Tools session with a single mono audio track. The track should contain an audio clip that has some unwanted noise in the background. Dialogue recorded in a room with an appliance running is a good example.

Make sure in your Pro Tools session that the **Inserts** column is shown in the Edit Window. You can do this by going to the menu bar and selecting **View | Edit Window Views** and ensuring the appropriate items are checked or by using the **Edit Window View selector** drop-down option directly above the track headers on the left side of the Edit Window.

You will also find having **Loop Playback** enabled under the **Options** menu helpful to loop over a selection as you listen for the changes.

How to do it...

We'll use one of the basic plugins that come with Pro Tools to reduce some of the noise in a track. Follow along with these steps:

1. Activate **Select Tool** and highlight a portion of an audio clip.

2. In the audio track header, click on the first available insert slot.

3. Select **plug-in | Dynamics | Channel Strip**.

4. Press the spacebar or click the play button in the transport to begin the playback.

5. In the middle of the **Channel Strip** plugin window, click on the **EXP/GATE** tab:

Figure 5.1: Avid Channel Strip

Figure 5.2: The EXP/GATE tab

6. Click and drag the **RATIO** knob up to 1 : 2 . 5 or click on the number next to the knob and type in the value.

7. Click and slowly drag the **THRESH** knob up until you hear the noise is reduced, but it doesn't affect the dialogue.

8. Click and drag the **KNEE** knob up to 14 dB or click on the number next to the knob and type in the value.

9. Click and drag the **RELEASE** knob down to 50 ms or click on the number next to the knob and type in the value.

10. Click and slowly drag the **DEPTH** knob up until you hear a balance between the noise and dialog that is acceptable to your ears.

11. Click the power button on the top left of the **EXP/GATE** tab to disable it and hear the difference.

12. Toggle the **EXP/GATE** power on and off to assess whether the end result is effective and make slight adjustments to the knobs as needed.

How it works...

There are several different approaches to noise reduction, and we're exploring one of the most basic types that still yields decent results. Expanders work by attempting to lower the volume of any part of the signal that's quieter than the threshold you set. You can imagine it like someone watching the levels on an audio signal and raising the volume knob whenever they see the signal go past a certain decibel level and lowering it when they don't. This type of noise reduction doesn't work in all scenarios and can leave unwanted frequencies present, but it's still useful for many different mix situations and can help when trying to bring up sounds while reducing the noise floor.

It's called an Expander because it tries to make the louder sounds louder and the quieter sounds quieter, hence "expanding" the dynamic range. When the Ratio of an expander is set to the maximum setting, then it functions as a Gate. Gates only allow sounds above the threshold through, creating a much "choppier" sound. This can be used with great effect in percussion but is not recommended for voice applications.

Pro Tools also comes with a dedicated Expander/Gate called Dyn3 Expander/Gate, but in my experience, Avid Channel Strip is better-sounding and has more features. If you're using a lower-tier version of Pro Tools, you may not have Avid Channel Strip available, so you would use Dyn3 instead. It can still perform well under certain situations and follows the same workflow.

The knobs we manipulated are as follows:

RATIO

This is how much the sound below the threshold will be reduced. The higher the number, the more aggressive it will be in lowering the quieter sounds. I usually opt for a higher threshold to start and roll it down if the reduction is too harsh.

THRESH (threshold)

This is the point of loudness in which the expander will allow sounds to pass through it freely. There is no correct threshold setting for any scenario, you need to listen and make a judgment call every time you use it. Both Channel Strip and Dyn3 offer a graphic display to represent the audio signal and how it's being affected by the expander. You can use this to help determine what the threshold should be. Find a spot in the audio where there is only noise and see where it "dances" on the graph. Your threshold should usually be set around there.

KNEE

You can see the effect the knee has on the graph. The top of the line represents the threshold of the gate. The default is to have a "hard knee", where the threshold appears like a single point. If you increase the knee value (a "soft knee"), the threshold becomes more curved. This has the effect of making the noise reduction more gradual and usually more natural-sounding.

RELEASE

The **RELEASE** knob sets how quickly the expander goes back into effect to lower the signal. When a release is set too quick (low), the result can be choppy and noticeable. When it is set too high, the noise continues to be heard after the signal stops. It might be subtle but can have a noticeable impact on the end result and the perceived noise reduction.

DEPTH

This allows you to bring up the overall noise floor (the part of the signal the expander is trying to lower). In most situations, having a completely dead signal cleaned of any noise can sound unnatural and unsettling to the audience. Therefore, many sound designers will reintroduce noise back into the mix but in a more controlled manner. Alternatively, you can reduce the depth to bring the noise floor up until it sounds more natural while still reducing it.

There are two knobs we didn't play with, **ATTACK** and **HYST**. Attack is how quickly, after crossing the threshold, the gate allows the sound through. When using an expander/gate to reduce noise, this should be set to a fast attack time, such as 20.0 us. **HYST** is short for hysteresis and is only available when the Ratio is set to its maximum, making it function like a Gate. Hysteresis is designed to help smooth the sounds of a harsh gate by allowing sounds below the threshold to pass through when on the rise.

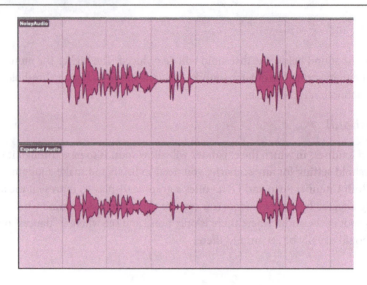

Figure 5.3: A waveform before and after applying an expander

There's more

Using an Expander/Gate is an effective way to reduce unwanted noise in the audio, but it can only go so far. There is a myriad of other noise reduction plugins that have different methods, price points, and pros and cons. While not an extensive list, here is an offering of some of the different noise reduction tools I've used over the years, along with my thoughts on them.

Waves Clarity

One of the newer offerings on this list, Waves Clarity uses machine learning and neural networks to clean up human voices. The results are quite impressive, cleaning up sounds effectively while retaining detail. There are two versions: Vx, its "One-Knob" solution, and Vx Pro, which offers more advanced features. While I've used this on almost every spoken word project since its release, it is designed for voice only, so cleaning up musical instruments or other sound effects is not possible. It will at times also mistake certain vocal noises such as huffs and snorts as noise and remove them. Vx is also fairly affordable compared to some of the other tools on this list.

iZotope Rx

RX is not one plugin but a suite of tools designed for audio repair. It also includes a standalone audio editor and an AudioSuite plugin to move audio back and forth between it (RX Connect). While its Voice De-noise is not quite as impressive as Waves Clarity, it's very capable and can do many things that Clarity cannot, including De-reverb, mouth De-click, and spectral editing, and the de-noising tools are not voice only, so it can be used in a variety of scenarios. There are different tiers and upgrade offerings, so you can start with the most basic option and move up over time as needed, but the higher tiered versions can be expensive.

Wave Arts MultiDynamics

This one is my "ace-in-the-hole," and I continue to use it in my daily operations. As a multiband expander/compressor, it offers a whole host of dynamics processing as well as a very transparent noise reducer. I often will use it in conjunction with other noise reduction tools to catch small sounds that the others let through. It also has a large number of presets for different scenarios, including many musical ones. It can be purchased on its own or as part of the PowerSuite and it's in the middle of the road in terms of cost. It's also very lean in terms of CPU usage, so it can be useful for those running not-so-powerful systems.

ReaFIR

Lots of people love Reaper as a **digital audio workstation** (**DAW**), and when it offers the ability to use its plugins outside its DAW for *free*, it's hard not to feel some affection for the team at Cockos. Much like Wave Arts MutiDynamics, ReaFIR is a multifunction dynamics processor, but where the former relies on standard expansion/compression, ReaFIR uses **Fast Fourier Transforms** (**FFTs**) to process dynamics. While beyond the scope of this book, FFT is the mathematical basis behind many audio tools, including spectrographs and signal recognition in machine learning. Using these tools, ReaFIR can build a noise profile from your track and reduce it. Note that this is a Windows-only tool and will not work with Pro Tools but can be used in other DAWs.

Audacity

An open source and free DAW, Audacity is popular amongst many audio enthusiasts wanting a simple tool to work with audio. I still use Audacity at times for quick processing work, and its Noise Reduction tool is quite capable. A noise profile needs to be captured first, so having a few seconds (or more) of just noise on the track helps.

Cedar Studio

The most expensive offering on this list, Cedar products are usually opted for high-end professional studios or forensic operations. The tools they provide operate with both software and hardware, and some audio engineers swear they are the most effective at noise reduction. Pricing is not publicly shown for most of their products, but they require a not insignificant investment.

Using EQ to shape the tone of a sound

Next to Volume Automation, EQ will most likely be one of your most often-used tools. The concept of EQ is simple enough; you raise and lower specific frequencies' dB levels within a sound to emphasize or reduce the bass, mids, treble tones, and everything in between, but where a lot of people struggle is homing in on which tones need to be adjusted. There's no easy way through it; good use of EQ takes practice, ear training, and time. However, it's extremely helpful to know the ins and outs of an EQ plugin as it will offer you some of the fundamental ways EQ should be used. It's also beneficial that the stock EQ plugins for Pro Tools are quite functional as well.

Getting ready

For this session, you will need a Pro Tools session with a single mono audio track. The track should contain an audio clip.

Make sure in your Pro Tools session that the **Inserts** column is shown in the Edit Window. You can do this by going to the menu bar and selecting **View | Edit Window Views** and making sure the appropriate items are checked or by using the **Edit Window View selector** drop-down option directly above the track headers on the left side of the Edit Window.

You will also find having **Loop Playback** enabled under the **Options** menu helpful to loop over a selection as you listen for the changes.

How to do it...

For this recipe, we'll take a track and play with some EQ settings to see the result. Follow along with these steps:

1. Activate **Select Tool** and highlight a portion of an audio clip.

2. In the audio track header, click on the first available insert slot.

3. Select **plug-ins | EQ | EQ3 7-Band**:

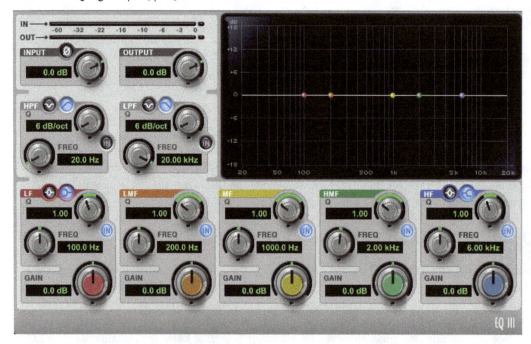

Figure 5.4: EQ3 7-Band

4. Press the spacebar or click the play button in the transport to begin the playback.

5. In the EQ plugin window, click and drag the red knob labeled **GAIN** on the bottom left in the **low frequency** (**LF**) area up and down while listening to the playback and observe the effect it has on the tone of the audio.

6. Click and drag the red LF's **GAIN** knob up to +12 dB, or click on the numbers displayed next to the knob.

7. In the **LF** area, click and drag the topmost knob up and down (it's above the **IN** button) and observe the change in the sound – this is the **Q** knob.

8. In the **LF** area, click and drag the leftmost knob labeled **FREQ** up and down and observe the effect it has on the tonal quality – this is the frequency knob.

9. Hold down *option* (*Alt* on Windows) and click on all three knobs in the **LF** area – **GAIN, FREQ,** and **Q** – to reset them to default.

10. Repeat *steps 6–10* in each of the colored areas and listen to how things sound as these settings are changed

11. Locate the **high pass filter** (**HPF**) area toward the middle left of the plugin window and click the **IN** button to activate it.

12. Click and drag the **HPF** area's **FREQ** knob up and hear how it removes the bass frequencies.

13. Click and drag the **HPF** area's **Q** knob (it's indexed in 6 dB increments) and hear how it removes more of the lower frequencies.

14. Click the **HPF** area **IN** button to disable it.

15. Repeat *steps 12–15* for the **low pass filter** (**LPF**) area and hear how it removes the higher frequencies.

How it works...

EQ3 7-Band EQ is a parametric EQ in that you can adjust specific parameters and make drastic changes to how something sounds. Conversely, other EQs might only offer more broad adjustments, such as only bass, mids, and treble, or you might have a Graphic EQ that functions with specific frequency bands across the spectrum (this is more common in live audio applications).

While you have more flexibility and control with a parametric EQ, there are still limitations with EQ3 7-Band EQ; specifically, the five bands at the bottom of the plugin all have a range for the frequencies they can adjust. You cannot increase the LF band to higher than 500 Hz, for example.

Let's break down the different functions for each knob type within the EQ to get a better understanding of what they do.

GAIN

The **GAIN** knob increases or decreases the volume of that specific band. It can go up to +12 dB and as low as -12 dB. The graph shown on the top right plots gain on the *y* axis, with the middle being 0 dB, positive values higher, and negative values lower. If you are listening to something and think it needs more bottom end or bass, you can try increasing the LF Band as an example. Conversely, if the track you are listening to has too much bass, you can lower the **GAIN** knob on the **LF** band to lower those bass frequencies.

FREQ (frequency)

This is the frequency that the specified band is targeting, measured in **hertz** (**Hz**) or the number of times a signal vibrates in a cycle per second. The graph displayed shows all the bands plotted along the *x* axis, with the lower frequencies (bass) on the left and higher frequencies (treble) on the right. The graph spreads across the range of human hearing, 20 Hz to 20,000 Hz (20 kHz). If you are listening to a track and find that it tonally doesn't sound like what you are aiming for, you can increase or decrease the gain on a band and then sweep the frequency left and right to determine where the change should occur. Over time you should begin to make associations with what you are hearing and the numerical value associated with its frequency, but even after years of practice, I still sweep frequency bands to help me lock into what the EQ should be targeting. Frequency can also be considered pitch.

Q

Q is how wide or narrow the band you are targeting is affected. A band with a high Q setting will only affect very few frequencies, sometimes only one. A band with a low Q setting will affect a much wider number of frequencies. For example, if I listen to a track and I hear a whine or hum, I can set one of the bands to a high Q and lower its gain to reduce how prevalent it is (I'd still have to identify the offending frequency of course). Alternatively, if I listen and hear that the sound is very muffled (perhaps it was recorded with a lavalier under a shirt), then you can use a band with a lower Q to affect a wider range of frequencies towards the high end and raise them to increase the clarity.

Other buttons

The **IN** buttons can be thought of as on/off switches. This was the term used on hardware to indicate that a signal was actively going *into* that module, so it is still prevalent in many plugins to this day. By default, **HPF** and **LPF** are both deactivated.

Both the **HPF** and **LPF** areas have two buttons toward the top that indicate the shape and style of filter they will behave as. The default is a "cut" filter that will cut off all frequencies above or below the point at which it is set; this is denoted by the slope graphic. The alternative option, which looks a bit like a "v" shape, is a notch filter, which will try to completely remove the frequency at which its located. When using notch mode, the Q value behaves like a standard **Q** knob, but in the cut mode, it's indexed to 6 dB/octave values and will make the slope steeper or shallower.

Both **LF** and **HF** also have two buttons toward the top of their areas. The default, which looks a bit like a tuning fork, is the **shelf** mode. This will evenly increase or decrease all frequencies above or below the set point. The other button, which looks like a circle, will turn those bands into standard EQ bands, similar to the other three in between them.

There's more...

EQ3 7-Band is not the only option for EQ, with many other third-party offerings and even other options within Pro Tools. Here's a brief list of other options that I've used over time. The important thing to consider when choosing an EQ is each one has its intricacies in terms of the **user interface** (**UI**) and functionality. Each tends to have its own audio characteristics in terms of how they sound – so it can be very subjective as to which one will work best for any specific scenario and to the engineer's taste.

When learning which frequencies to address in a mix, a frequency analyzer can also help immensely. Some of the tools listed include one, but you can also purchase plugins that work as standalone analyzers. These will plot the frequencies along a graph from low to high. You can then compare what you see with what you hear and adjust the EQ accordingly.

EQ3 1-Band

When you only need one band to be affected, such as only an HPF, or only notching out one frequency. There's only one area to worry about, but it has six different types, all of which were discussed previously.

Channel Strip

We explored this in the previous recipe, but the Channel Strip functions as a multi-tool of sorts combining dynamics processing and EQ. There is a very capable EQ built into it, although it has bands as opposed to 7, and is very compact, so a bit clunkier to use compared to EQ3.

iZotope Neutron

Functioning more like a Channel Strip by combining dynamic processing with an EQ module, the UI provides a lot of great feedback, like a frequency analyzer, to show you how your changes are affecting the sound visually. It also comes with a myriad of presets to help you out.

Waves Q10

One of the staples of many post-production houses, Q10 offers 10 bands of EQ control and does not limit which frequencies they can be applied to. With the ability to manipulate EQ curves very accurately, it's possible to get great bandwidth-limited sound effects with it, such as the sound of someone speaking over a radio or telephone. It's also quite prevalent due to it being included in many of the Waves bundles.

FabFilter Pro-Q

One of the best EQ's I've used, Pro-Q offers up many bands of EQ, a visual analyzer, and Dynamic EQ, which essentially turns it into a multiband compressor. It includes many others such as EQ Match to help you match sounds and even the ability to input MIDI controls; it's definitely worth the higher price (as are all of the offerings from FabFilter).

How to clean up hum and crackle with Notch Filters

Sometimes the audio you are working with has specific tones or noises that are consistent across the track. These could be pops and crackles caused by electrical or radio inference, frequencies that resonate through a space, or even equipment and lighting in the area. Sometimes, even handling microphones can cause these unwanted noises. With some effort, it is possible to use just an EQ setup in a specific way to remove these offending frequencies. We'll use the EQ3 1-Band to both detect and eliminate unwanted noises from a track.

Getting ready...

For this recipe, you will need a Pro Tools session with a single track of audio (mono or stereo). While the effect can be heard in almost any piece of audio, finding one with an offending hum or tone to it will yield the best results. Recording a voice near a lighting fixture or other electrical equipment such as a refrigerator or furnace/air conditioner is a good example.

Make sure in your Pro Tools session that the **Inserts** column is shown in the Edit Window. You can do this by going to the menu bar and selecting **View | Edit Window Views** and making sure the appropriate items are checked or by using the **Edit Window View selector** drop-down option directly above the track headers on the left side of the Edit Window.

You will also find having **Loop Playback** enabled under the **Options** menu helpful to loop over a selection as you listen for the changes.

How to do it...

We'll add a single band EQ with a high Q and gain value, sweeping the frequency to identify the offending frequency, and then dropping its gain to remove it, essentially creating a notch filter. Follow along with these steps:

1. Click on the first available insert slot and select **plug-in | EQ | EQ3 1-band**.

2. In the middle of the plugin window, click and drag the **FREQ** knob all the way to the left so it displays 20.0 Hz, or click on the number value and type it in.

3. Above the **FREQ** knob, click and drag the **Q** knob up to 10.00, or click on the number value and type it in:

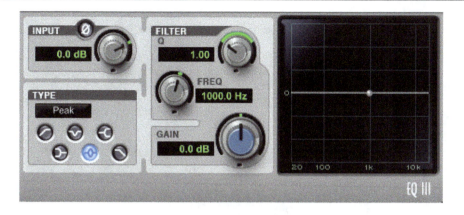

Figure 5.5: EQ3 1-Band

4.   Start playback with the spacebar or by pressing play on the transport.

> **Warning**
>
> These next steps can be harmful to your hearing if you're listening with loud playback levels. It might be helpful to temporarily turn down the volume on your monitors/headphones while trying to find the offending frequency.

5.   In the bottom middle of the plugin window, click and drag the **GAIN** knob to 18.0 dB, or click on the number value and type it in.

6.   Click and drag the middle **FREQ** knob moving it slowly to the right.

7.   Listen closely as you sweep the frequency until you hear the specific tone you are hoping to remove become emphasized; it should be very noticeable when you hit it (like a loud ringing in the audio); stop moving the knob when you do.

> **A note on sweeping frequencies**
>
> Sweeping frequencies to find them can be challenging, depending on the material. You may have to start again or move it back and forth to land on the exact sound you are trying to remove. Depending on your mouse sensitivity, you may also have to lift and move your mouse to reset its position several times to sweep all the frequencies. When sweeping, it helps to consider which tone you hear currently emphasized compared to what you are targeting and try to move the frequency toward that.

8.   Once you have identified the offending frequency, you can either decrease **GAIN** to -18.0 dB or change **TYPE** for the EQ in the bottom left of the plugin window to **Notch**, which is the top middle button that looks like a "v" shape:

Figure 5.6: Using EQ3 1-band as a notch filter

9. If there are hums or offending tones still present, you will need to add another instance of EQ3 1-band, so repeat *steps 1–8*.

How it works...

When applying a notch filter, you might think that completely muting a frequency would harm the audio quality. While it's true that you are removing audio information, since the scope of what you are removing is so small (just one frequency and possibly a few to the side of it), the result is rarely damaging. In some unique situations, you may find that removing it entirely is detrimental, and in those cases, simply increase the gain slightly to your satisfaction.

Some common use cases for notch filters and their range of frequencies are as follows:

*   Electrical equipment and lighting tend to hum at the frequency of the electrical standard of their region. In North America, this is **60 Hz**, while in Europe, this is **50 Hz**.

*   Rooms that resonate with voices tend to fall within the range of human voices, **200-2,000 Hz**; this often sounds like a hum or ringing that decays quickly.

*   Crackle and pops tend to register much higher, in the **3-8 KHz** range.

There's more...

You will also often find that there are harmonics of the frequency you are trying to eliminate present. These are fortunately relatively easy to remove; simply multiply the offending frequency by a value of 2, 3, and sometimes 4 or 5 depending on how prevalent the harmonics sound. For example, if you are removing an electrical hum at 60 Hz, then you will most likely also need to apply a notch filter at 120 Hz (60 x 2) and 180 Hz (60 x 3). These harmonics are usually not as loud as the target frequency, so the gain can be less aggressive with them.

Using Pro Tools' EQ3 to notch out several harmonics means having to use more inserts, so sometimes this task is best left to an Aux track to free up inserts on the target audio track. It's also sometimes best to look at alternative plugins to tackle this task. For example, Waves Q10 allows every frequency band to work as a notch filter across the spectrum, so one instance of it is enough to tackle multiple harmonics. There are also dedicated de-hum and de-crackle plugins such as in iZotope RX. These can typically automatically determine which frequencies and harmonics need to be removed.

Using Reverb to create space

Reverberation (Reverb) is a key component of any post-audio workflow. We don't live in anechoic chambers. Every space we experience has varying levels of reverb present as sound waves reflect, disperse, and decay when they interact with physical objects. Some of these spaces are engineered with reverb in mind, such as churches and concert halls, and some are just a product of the functionality we require from them, such as a tiled bathroom. When applying reverb to a track, your goal can be to add depth, richness, and "air" to a track, and sometimes it's to add realism to match other recordings. Whichever the case, the methods of routing audio to an aux track and applying the reverb there is the same, and we'll use that method in this recipe.

Getting ready

For this session, you will need a Pro Tools session with a single audio track. The track should contain an audio clip that you want to add reverb to.

Make sure in your Pro Tools session that the **Inserts** and **Sends** columns are shown in the Edit Window. You can do this by going to the menu bar and selecting **View | Edit Window Views** and making sure the appropriate items are checked or by using the **Edit Window View selector** drop-down option directly above the track headers on the left side of the Edit Window.

You will also find having **Loop Playback** enabled under the **Options** menu helpful to loop over a selection as you listen for the changes.

How to do it...

We'll send an audio track's signal to an aux track and apply reverb to it there. You can do this with these steps:

1. Click on the first available **Send** slot and select **new track...**.
2. Set **New Track** to the following values:
   - **Format**: Stereo
   - **Type**: Aux Input
   - **Time Base**: Samples
   - **Name**: Reverb

3. Click **Create**.

4. In the newly created Reverb Aux Track, click on the first insert available and select **multichannel plugin** | **Reverb** | **D-Verb**:

Figure 5.7: D-Verb

5. Make sure the send channel's fader is active and visible on the screen – if you don't see it, click on the Send named **Reverb** in the **Send** column.

6. Start playback with the spacebar or click the play button on the transport.

7. Click and drag the send level's fader up slowly and listen to how it sounds with the audio being sent to the reverb track; stop when you like how it sounds.

8. While the track is playing back, experiment with different presets in the D-Verb plugin window – you can do this by pressing the different buttons at the top that read **Hall**, **Church**, and so on or by clicking on the top drop-down menu in the **Preset** area that reads **<factory default>** and select the different presets there to try out.

How it works...

It is possible to simply add reverb as an insert to an audio track, but the result typically sounds like a mic that's pointed away from the source of the sound and picks up only the sound of the reverb in the room. Most reverb systems come with a wet/dry knob or slider (as does D-Verb), but this still doesn't give the same effect that would be recorded should someone record with a mic close to the source in a space. By using a send, you get that nice blend of presence from the main audio channel combined with the reverb mixed in the amount you want it. Being able to control not only how much of the signal gets sent to the reverb channel but also how loud the reverb itself is (by adjusting its volume) results in much more natural-ounding reverbs in a mix.

This method also allows you to send multiple tracks to the same reverb channel, giving a consistent sound to everything. For example, if a band was performing in a room together, you would expect they would have a similar-sounding reverb.

One small note, notice that we selected **multichannel plug-in** as opposed to **multi-mono plug-in** when selecting the insert. These options are only available on stereo tracks. Multi-mono plugins process the left and right channels independently, while multichannel ones work together. In the case of reverbs, this means that sounds that emanate from the right channel reflect differently on the left and vice versa, typically giving a much more natural sound. Some plugins are only available in one form or another; D-Verb is present in both. There may be some rare situations where you would want multi-mono reverb, but 99% of the time, you will find the results better in multichannel.

There's more...

It's sometimes worthwhile combining different reverbs together. You can do this by using separate sends to separate Aux tracks or simply setting multiple Aux tracks to the same input through the tracks I/O column.

D-Verb is a very capable reverb plugin, and I use it often in musical situations, but for motion picture and spoken word, it's nice to have reverb plugins that are designed to emulate rooms more accurately. Here is a small sample of other reverb plugins currently on the market that you can try out. Each has its own characteristics in terms of the reverbs it yields and the interface to control them, so it's worth taking advantage of a free trial to see whether they suit your needs:

- Waves R-Verb
- FabFilter Pro-R
- Valhalla Room, Vintage Verb, and Supermassive
- denise Perfect Room
- Altiverb (see the following note)

> **Convolution reverb**
>
> While it might sound complicated, the practice of using convolution reverb is quite simple. You find a space and record a frequency sweep using good speakers and microphones, or you record an impact such as a hand clapping, a balloon popping, or two bricks being hit together. These create an **impulse response (IR)**, which can then be used in a convolution reverb plugin to recreate the sound of that space. Alitverb is a convolution reverb and comes with a myriad of IRs. You can also find free IR plugins such as Convology XT by Impulse Record to use your own IR recordings.

Making things strange and scary with reverse reverb

You have surely heard this effect before. It might be hard to describe and place exactly where, but it's like the sound is swelling up into the track before it plays. While the origin of it is contested, the practice was originally performed with analog tape, flipping the direction of the tape to reverse the playback, recording a reverb track of that, then flipping both tracks. What you get is the sound of the reverb building up to the start of the track. It can be unnerving and offsetting or trippy and mystical. However you describe it, it's a great technique to have in your arsenal as an audio designer.

Getting ready

For this recipe, you will need a Pro Tools session with two mono audio tracks. One track should have an audio clip. While any sound can work, this effect is used most often in musical and dialogue situations. This recipe also uses hotkeys, so either make sure the edit window hotkeys button is active (small box with the letters **az** on the top right of the timeline) or use *control* (the Windows key on Windows) as a modifier key in conjunction with the key commands.

How to do it...

We will duplicate a track, reverse it, print reverb to it, then reverse it back to create the effect. Follow along with these steps:

1. Use **Grabber Tool** to move the audio clip forward in the timeline so that it has about 5 seconds of lead-in time.

2. Use **Select Tool** and highlight the portion of the track you wish to apply the effect to – if it's a short clip, you can select the entire of it.

3. Press *c* to copy the clip.

4. Move the selection to the next track down with the ; key.

5. Press *v* to paste the clip.

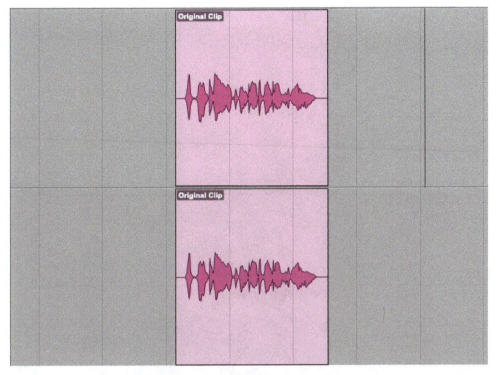

Figure 5.8: Duplicating a clip to a new track

6. Use **Select Tool** to highlight a large area around the newly pasted clip; about 5 seconds before and after it.

7. In the menu bar, select **AudioSuite | Other | Reverse**.

8. Make sure the second drop-down menu at the top left below the **Reverse** drop-down option is set to **create continuous file**.

9.   Click the **Render** button at the bottom right of the plugin window:

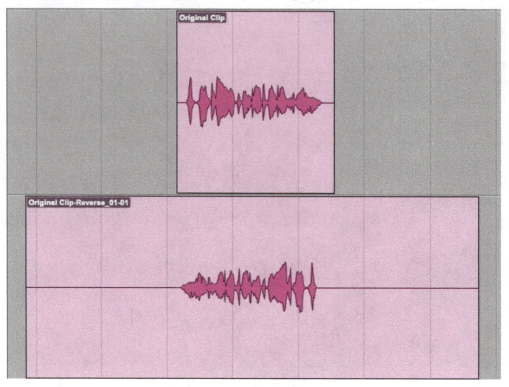

Figure 5.9: The clip has been reversed

10. In the menu bar, select **AudioSuite | Reverb | D-Verb**.

11. Click the button in the top middle of the D-Verb plugin window that says **Church**.

12. Click the **Render** button on the bottom right of the plugin window:

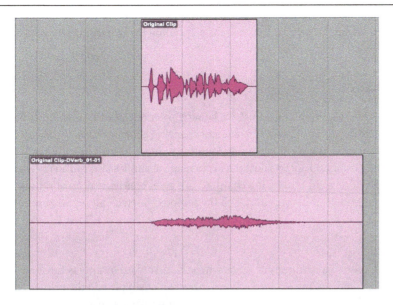

Figure 5.10: Reverb has been printed on the clip

13. In the menu bar, select **AudioSuite | Other | Reverse**.

14. Click the **Render** button at the bottom right of the plugin window:

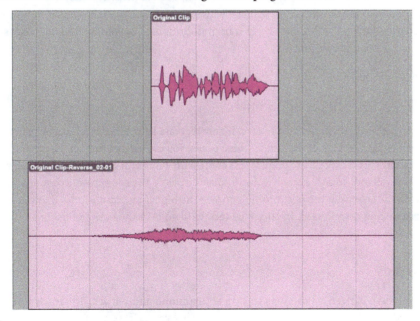

Figure 5.11: The reverb clip has been reversed once more, creating the reverse reverb effect

15. Start playback with the spacebar or press the play button in the transport to listen to both tracks playing together and hear the effect.

How it works...

This effect relies on the selection made in *step 6* being long enough both before and after the pasted clip to accommodate the decay of the reverb and not changing going forward. If you accidentally click somewhere else afterward, it's not a big deal so long as the reverse effect has already been rendered. You can simply use **Grabber Tool** to select the same range again. It's also fun to try and experiment with different types of reverbs. I find **Church** in D-Verb gives consistently great results, but you can try out the different presets or play around with the settings yourself and see what results you get.

There's more...

This effect is often used in a manner that stops when the first word or note is finished. You can edit the audio clip before applying the effect to achieve this effect or simply edit the printed effect after the fact. It's also fun to experiment with non-reverb effects in a similar manner to see what the results are.

Using compression to control dynamics

A compressor works like someone riding the volume fader. You tell them at what point of loudness to turn down the volume (the threshold), how much you want to turn it down by (ratio), how quickly to apply the reduction (attack), and how long to wait until bringing the volume back up (release). The use case for compression is two-fold:

- To reduce the peaks and even out the loudness of a track
- To bring the overall volume up if needed

If a track has no compressor applied, then you can only increase its volume as far as the peak loudness will allow before it clips/distorts. By squashing these peaks, you can safely bring up the overall loudness without distortion. This is why most compressors come with a makeup gain parameter. Compressors also affect the overall tone of a track, either due to their circuitry, the components they are modeling, or simply due to the change in the relationship between the higher and lower tones that were affected by the dynamics being compressed. In later chapters, we'll go over the basics of using a compressor with specific use cases for different scenarios.

Getting ready

For this recipe, you will need a Pro Tools session with a mono audio track. The track will need an audio clip placed in it. This clip should be one you recorded yourself, as most pre-recorded loops or samples tend to be compressed already.

Make sure in your Pro Tools session that the **Inserts** column is shown in the Edit Window. You can do this by going to the menu bar and selecting **View | Edit Window Views** and making sure the appropriate items are checked or by using the **Edit Window View selector** drop-down option directly above the track headers on the left side of the Edit Window.

You will also find having **Loop Playback** enabled under the **Options** menu helpful to loop over a selection as you listen for the changes.

How to do it...

We will place one of Pro Tools' built-in compressors on a track's insert and manipulate the parameters. Our goal is to determine what settings work best for this particular track, as all tracks will have different needs. Follow along with these steps:

1. In your track header's **Insert** column, click on the first available insert slot and select **plug-in | Dynamics | Dyn3 Compressor/Limiter**:

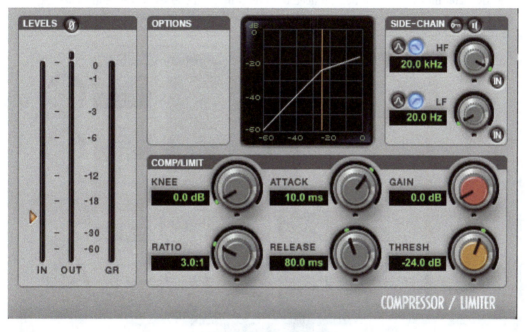

Figure 5.12: Dyn3 Compressor/Limiter

2. In the bottom area of the plugin window, click and drag the **RATIO** knob and increase it to its maximum value, 100:1 – you can also click on the number and type it in.

3. In the same area, click and drag the **ATTACK** knob to its lowest value, 10.0 us, or click the number and type in the value.

4. Directly below, click and drag the **RELEASE** knob to its lowest value, 5.0 ms, or click the number and type in the value.

5. To the right of that knob, click and drag the orange **THRESH** knob to its highest value, 0.0 dB, or click the number and type in the value.

6. Begin playback with the spacebar or click the play button on the transport – the next step does not require you to listen to the playback; you will be monitoring the meters visually.

7. Click and drag the orange **THRESH** knob and slowly turn it down; as you do, watch the third meter from the left and see how it reacts.

8. Continue watching the **Gain Reduction (GR)** meter as you lower the knob and stop reducing the threshold when you see the meter is most active

> **Finding the threshold**
>
> Determining the correct threshold can take a bit of playing around to get correct. If you see the **GR** meter moving just a little, you need to lower it more. If you see it constantly lowering (see *Figure 5.13*), it's too low, and you need to bring the threshold back up.

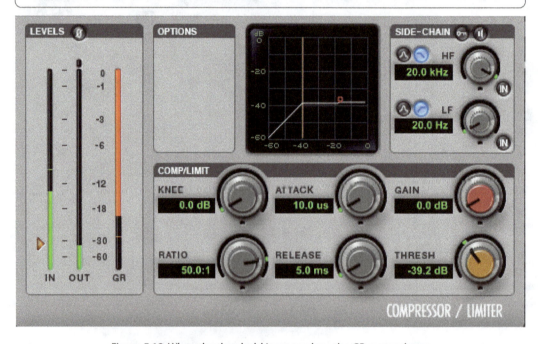

Figure 5.13: When the threshold is set too low, the GR meter shows
that too much compression is being applied

9. Once you've found the ideal threshold, you can stop playback.

10. Hold *option* (*Alt* on Windows), and click on the **RATIO, ATTACK**, and **RELEASE** knobs to reset them to their default values.

11. Start playback again.

12. Click and drag the **RATIO** knob up to a higher amount; you can play with different values and hear the different results – try ratios such as 3 : 1, 4 : 1, 5 : 1, and 8 : 1.

13. Click and slowly drag the red **GAIN** knob up to increase the overall loudness.

14. While still playing back the audio, click the **BYPASS** button on the top middle of the plugin window on and off to toggle the effect and hear how it sounds; you can also *command* + click (*Ctrl* + click on Windows) on the insert in **Edit Window** to enable or disable the bypass.

How it works...

The biggest challenge when using a compressor is finding the correct threshold that will provide the most impact without overdoing it.

By increasing the ratio and decreasing the attack and release time, you are only playing with the threshold as a variable. Every single audio file you work with will have different audio levels, so sometimes, you will need to have the threshold very low (for example, -30 dB or even lower), or sometimes it will be quite high depending on the material (-12 dB or higher). If a threshold is too high for a specific track, then the compressor is not really providing the full effect it can. Conversely, if it's too low, it is constantly active and not really compressing the dynamic range of the sound so much as just lowering the volume overall.

There are meters and graphs to help you determine the correct threshold as well, with a convenient orange triangle on the left of the **IN** meter that coincides with the threshold level and a graph with a red dot on the screen to show where the levels are at any moment. The graph represents the input/output relationship, with the x axis being the input sound and the y axis is the output. Notice how the white line starts to slant to the right of the threshold line (also in orange) and how steeply it corresponds with the ratio. This is trying to illustrate that the sounds that go above that threshold line will be lowered accordingly:

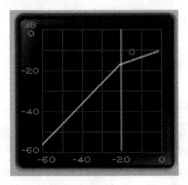

Figure 5.14: The compression graph with the peak of the signal indicated by a red circle

You can also use the white line on the graph as an indicator of the output's loudness when the gain is applied. If you only compress the sound, then it will sound quieter overall compared to the original source. Using the gain will bring up the overall loudness, and you will see the line in the graph move up accordingly. Be careful, though, as this also brings up the noise floor, so some care needs to be applied when setting both threshold, ratio, and gain.

Finally, the right attack and release settings are highly dependent on the sound the compressor is being applied to. Things that tend to have a longer decay, such as bass instruments (depending on the genre) usually require a longer release time. Attempting to apply a compressor with too quick an attack to percussion tracks will lose all the transients needed for the strong impacts. Using the presets available is always a good way to try different settings and see how they affect the sound.

There's more...

Dyn3 Compressor/Limiter is a good way to learn how compressors work, and I still use it for specific situations, but I don't find it the best-sounding compressor overall. Avid Channel Strip includes an excellent-sounding compressor, but there is a multitude of other compressors on the market that have distinct characteristics and tonal qualities. Talk with other engineers in the field you work in and download some free trials to see which ones work best in the medium you are using.

Limiters also function very similarly to compressors, hence the fact that Dyn3's offering is named a Compressor/Limiter. They essentially are a compressor with a very fast attack and release and a very high ratio. Limiters typically are used to increase the overall loudness of a track without affecting the tone as much.

Making sounds louder without killing transients with parallel compression

Compression is a great tool to level dynamics and makes things sound louder, but as noted in the previous recipe, in certain situations, you risk losing the transients or first moments of a sound's attack. For certain instruments, such as percussion, guitars, or pre-mixed beds of music, this can absolutely kill the sound and make it sound dead in comparison. In situations like this, you can use a technique called parallel compression, sometimes also referred to as upward compression. The concept is simple; instead of using a compressor to reduce the peaks of a sound and make up the gain after, you heavily compress a duplicate track (or, in our case, a Send routed to an Aux track) and mix that with the original track. The result is the lowest sounds being brought up without squashing the transients of the original.

Getting ready

For this recipe, you will need a Pro Tools session with an audio track. The track will need an audio clip placed in it. This clip can be from any source, but I find audio tracks that are musical in nature tend to show off this effect the best.

Make sure in your Pro Tools session that the **Inserts** and **Sends** columns are shown in the Edit Window. You can do this by going to the menu bar and selecting **View | Edit Window Views** and making sure the appropriate items are checked or by using the **Edit Window View selector** drop-down option directly above the track headers on the left side of the Edit Window.

You will also find having **Loop Playback** enabled under the **Options** menu helpful to loop over a selection as you listen for the changes.

How to do it...

We will send our track to an Aux track and apply some aggressive compression to it, then bring it into the mix to add some volume without reducing the transients. Follow along with these step:.

1. In your track header's Sends column, click on the first available send and select **new track…**.

2. Keep the default settings (Stereo, Aux Input, Samples) and name it COMP.

3. Click the **Create** button.

4. In the send fader window, hold *option* (*Alt* on Windows) and click on the fader slider to set it to 0 dB.

5. In the newly created **COMP** track, click on the first insert slot and select **plug-ins | Dynamics | Dyn3 Compressor/Limiter**.

6. At the bottom of the plugin window, click and drag the **RATIO** knob up to 50:1, or click on the number and type in the value.

7. Click and drag the **THRESH** knob down to about the lowest peaks of the **IN** meter on the left – this will be quite low, most likely below -20 dB.

8. Click and slowly drag the red **GAIN** knob up until it reaches a volume that pleases your ear; it should probably only need a few dB:

Figure 5.15: A typical parallel compression setup

How it works...

Parallel compression works by taking advantage of a fundamental property of sound – when two matching waves are added together, they increase in amplitude. The duplicated signal has an aggressive compressor that levels out the peaks to almost match the lowest peaks of the sound. When you add that back into the mix, there is almost no effect on those existing louder parts; there is already so much energy present in them. But the lowest parts have their peaks increased significantly in comparison. This retains the transients and punch of the audio while still giving it some loudness and some of the characteristics of the compressor.

There's more...

There are dedicated plugins that offer upward compression as an alternative. While there are many available, one of the most well known in audio engineering networks is called **Over The Top** (**OTT**). OTT is a free plugin that provides parameters for both downward and upward compression. When used lightly, it can add that extra bit of loudness and character to a track. However, when used too much, it can still provide some very interesting results for sound design.

Pitch shifting in real time with plugins

Pitch shifting as a musical practice has divided musicians and listeners alike when used as pitch correction. The use of Auto-Tune and other pitch correction plugins can be considered at times as both a creative tool and a crutch. This kind of pitch correction software does not come with Pro Tools. While the pitch-shifting tools that do come with it are more limiting, they are, at heart, designed to be used in broader terms and can be more functional. We'll first explore changing the pitch of a track in real time with a plugin.

Getting ready

For this recipe, you will need a Pro Tools session with a mono audio track. The track will need an audio clip placed in it.

Make sure in your Pro Tools session that the **Inserts** column is shown in the Edit Window. You can do this by going to the menu bar and selecting **View** | **Edit Window Views** and making sure the appropriate items are checked or by using the **Edit Window View selector** drop-down option directly above the track headers on the left side of the Edit Window.

You will also find having **Loop Playback** enabled under the **Options** menu helpful to loop over a selection as you listen for the changes.

How to do it...

We'll be adding one of Pro Tool's built-in pitch-shifting plugins, Pitch II, and playing with its parameters. Follow along with these steps:

1. In the **Insert** column in your track header, click on the first available insert slot and select **plug-in | Pitch Shift | Pitch II (mono)**:

Figure 5.16: Pitch II (mono)

2. Start playback with the spacebar, or click the play button in the transport.

3. In the middle left of the plugin window, click and drag the **COARSE** knob down to -1 semi, or click on the number and type in the value. Observe the keyboard has illuminated the key directly to the left of the key in the middle (middle C, which is now slightly lighter than the other keys).

4. On the keyboard, click on the key directly to the right of middle C and observe how the **COARSE** knob now reads 2 semi.

5. Click and drag the **FINE** knob up to its maximum `50 cents`, or click on the number and type in the value.

6. Click and drag the **FINE** knob down to its minimum `-50 cents`, or click on the number and type in the value.

7. Experiment with different **COARSE** and **FINE** adjustments and listen to the result – you can also play with the effects pane below, which can add different amounts of delay to the signal and mix in the original signal with the pitched signal on (wet/dry).

How it works...

Pitch Shift works by slicing up the signal into very small segments (6 ms to 42 ms), applying speed changes to those segments, and restructuring the signal back to recreate the same timing as the original. The result is a signal that sounds like it's been sped up or down with a traditional playback system, but the timing has not been affected. This has its limits, of course, and too extreme of a shift in either direction will result in noticeable artifacts.

The Transient pane is there to offer some help to fine-tune the result should you start hearing "robotic" or "stretched-out" sounds. Pitch Shift attempts to consider transients when processing the signal. If you start hearing issues with longer notes in music scenarios, you can try raising the threshold to fix this. Conversely, if you notice sharp attacks getting lost, try lowering the threshold. The window option is similar; it defines the length of the segments it uses in its processing. Sounds with very quick transients tend to benefit from a shorter window setting, whereas sounds that are becoming too robotic might be helped by a longer window.

The coarse and fine adjustments should be familiar to anyone with some understanding of music theory, but if you're not familiar: **Coarse** refers to semitones, which are one key apart on a piano (the difference between a B and C note, or a C and C sharp); **Fine** is the space between those notes, separated into 100 cents. Due to the way that notes are laid out, the fine control only goes up or down 50 cents. If you want to go beyond that, you'll need to move to the next semitone, up or down. For example, to lower a signal by 51 cents, you'd drop the **COURSE** knob -1 semitone and increase the **FINE** knob to 49 cents.

If the musical notes and percentages aren't easy to work with, simply ignore them and use the **RATIO** knob instead. Turn it up to increase the pitch and down to lower it.

You can also automate any of these plugin parameters to try shifting the pitch of a sound over time but don't expect Auto-Tune-like results. Due to the way Pitch Shift II works, the changes in pitch will be far less transparent and more difficult to manipulate.

Printing Pitch Shift effects with AudioSuite

While being able to manipulate the pitch of a sound in real time is useful for a lot of situations, I find Pro Tool's AudioSuite offerings in the pitch-shifting realm much more practical. In most scenarios, I want to have a pitch effect made permanent and printed, not changed in real time. An example is when trying to create variation with sound effects and foley in motion pictures. Sometimes, successive or repetitive impacts and certain sounds feel mechanical or artificial. By alternating the pitch of a sound effect as it repeats, the result is much more pleasant and "natural" to the audience. You can use pitch shifting to imply increases or decreases in momentum in a character or use it to match tones musically. Beyond using Pitch Shift II in AudioSuite to increase or decrease pitch without affecting timing, you also have two other powerful plugins that can be used – **Time Shift** and **Vari-Fi**.

Getting ready

For this recipe, you will need a Pro Tools session with a single mono audio track. Since we'll be using different effects several times, a short clip (about 3-5 seconds) that has been duplicated is best to have on the track. Three copies with a few seconds separating them will work best.

How to do it...

We're going to take a clip of audio and manipulate it in three different ways: alter its timing without affecting its pitch, alter both its timing and pitch, and create a speeding up or speeding down effect. Follow along with these steps:

1. Use **Grabber Tool** and select the clip on the track.

2. In the menu bar, select **AudioSuite** | **Pitch Shift** | **Time Shift**:

Figure 5.17: Time Shift

3. In the top area of the plugin window under **Audio**, click on the **MODE** drop-down option and
    select one of the following modes according to the type of audio you are using:

    - **Monophonic**: This is a single source of audio that doesn't vary drastically in tonality – single-
      voice instruments such as woodwind and brass instruments usually work well in this mode.

    - **Polyphonic**: This is a more complex source of audio with multiple tones or instruments –
      dialogue and music beds tend to work best in this mode.

    - **Rhythmic**: This is for percussion or instruments with regular timing impacts and transients.

4. On the right middle of the plugin window, click and drag the **SPEED** knob up to 150%, or click on the number and type in the value.

5. Click on the speaker at the bottom left of the plugin window to preview the result.

6. Click and drag the **SPEED** knob down to 75%, or click on the number and type in the value.

7. Click on the speaker at the bottom left of the plugin window to preview the result (click it again to stop).

8. Click the **Render** button on the bottom right to print the effect – this changes the timing of the track without affecting the pitch.

9. Use the **Grabber** tool to select the second clip on your track.

10. In the **Time Shift** plugin window, use the **MODE** drop-down option under the **AUDIO** area at the top and change the value to **Varispeed**.

11. In the right middle of the plugin window, click and drag the **SPEED** knob up to 150%, or click on the number and type in the value.

12. Click on the speaker on the bottom left of the plugin window to preview the result (click it again to stop).

13. Click and drag the **SPEED** knob down to 75%, or click on the number and type in the value.

14. Click on the speaker on the bottom left of the plugin window to preview the result.

15. Click the **Render** button at the bottom right to print the effect – this changes both the timing of the track and the pitch as if slowing down or speeding up an analog recording.

16. Use the **Grabber** tool and select the third clip in your track.

17. Close the **Time Shift** plugin window, and in the menu bar, go to **AudioSuite | Pitch Shift | Vari-Fi**:

Figure 5.18: Vari-Fi

18. Leave the default settings as follows:

    - **CHANGE: SLOW DOWN**

    - **SELECTION: FIT TO**

    - **FADES: ON**

19. Click on the speaker on the bottom left of the plugin window to preview the result (click it again to stop).

20. Change the **SELECTION** option to **EXTEND** and preview the result.

21. Change the **CHANGE** option to SPEED UP and preview the result.

22. Change the **SELECTION** option to FIT TO and preview the result.

23. Click the **Render** button at the bottom right.

How it works...

AudioSuite Pitch Shift effects print the effect to the clip(s) selected in the timeline. Because time is often a factor, keep this in mind, as a clip can collide with or overwrite another clip next to it when rendered. There are three main Pitch Shifting tools we explored – time stretch with no pitch shift, time stretch with pitch shift, and Vari-Fi, which speeds a clip up or down over time as if you are changing a tape's playback speed as it plays.

Time Shift uses a similar method to Pitch Shift II for its processing, so if you want to learn about how the Transient options work, check out the previous recipe. Time Shift also offers the ability to change the pitch of the clip on the bottom right, but its interface and options are more limited.

It's also possible to use Pitch Shift II as an AudioSuite plugin to print effects similar to the real-time effect. You can follow the same steps as in the previous recipe but select it from the AudioSuite menu instead of on an Insert.

6

Finishing a Project and Creating Deliverables

No matter how good your edit and mix are, it won't mean anything if you aren't able to deliver the product to the client's specifications. In the past, broadcast standards made it somewhat more predictable and consistent in terms of what was expected of a mix and its targets. In the last decade, however, with the move to digital streaming services becoming common, standards can vary greatly, depending on what platform the project is being uploaded to. Being able to hit those targets and pass quality assurance is vital for any audio engineer, and we'll be going over a few ways to accomplish that in this chapter.

The recipes we'll be going over are as follows:

- Bouncing a project to a finished file
- Printing directly to a track
- Making changes to a Print track
- Printing stems
- Understanding **Loudness Units Full Scale** (LUFS)
- Using Limiters for quality assurance and compliance
- Consolidating and archiving a project
- Creating templates

Technical requirements

The recipes in this chapter require at least Pro Tools Studio, though some workflows may work with Pro Tools Artist.

The Example Sessions and Audio Files for each recipe can be found at `https://github.com/PacktPublishing/The-Pro-Tools-2023-Post-Audio-Cookbook/`.

Bouncing a project to a finished file

What might be the most direct way to finish and export a file, **bouncing** as a term was coined during the analog days since the signal would "bounce" around different devices to end up on the record track. Keeping this phrase might seem outdated, but it also harkens to a time when bouncing in Pro Tools could only be done in real time to take advantage of its outboard gear. Now, there are different ways to bounce a track, but we'll look at the most basic options and weigh up their pros and cons.

Getting ready

For this recipe, you will need a Pro Tools session with a few tracks and some audio clips present on the track.

How to do it...

We're going to set up a session to bounce to a stereo audio file in both real time and offline mode. Follow along with these steps:

1. In the **Edit Window** area, use **Select Tool** and highlight the portion of the project you'd like to export – you can click and drag along a track/clip, or within one of the rulers.

2. In the **Menu** bar, go to **File | Bounce Mix...**:

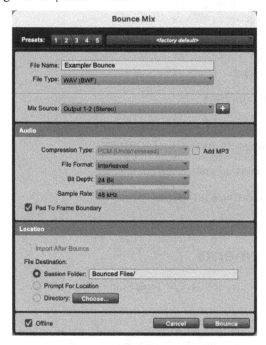

Figure 6.1: The Bounce Mix window

3. Type in a name you'd like to call the file in the **File Name** field.

4. Under **File Type**, select **WAV (BWF)**.

5. If you're using Pro Tools Ultimate, leave **Mix Source** at its default, which is **Output 1-2 (Stereo)**.

6. Under the **Audio** tab, select the following settings:

    *   **File Format: Interleaved**

    *   **Bit Depth: 24 Bit**

    *   **Sample Rate**: 48 kHz

    *   **Pad to Frame Boundary**: Off (unchecked)

7. Under **Location**, select **Session Folder** and enter Bounced Files/.

8. Make sure the **Offline** checkbox is unchecked in the bottom-left corner.

9. Press the **Bounce** button and listen to the track as it exports.

How it works...

Before bouncing a mix, it's always good practice to select the bounds of what you want to be exported; otherwise, Pro Tools will attempt to export what it thinks is the proper end point of your project, which means it will wait for no audio to be processed by any track or plugin. This can sometimes lead to exports being far longer than you expected, which can also be embarrassing if you leave a clip at the end of the session that you forgot to delete (this has happened to me several times). In some cases, Pro Tools might even cut off a track earlier than it needed to. Always use **Select Tool** to set your in and out points to avoid surprises.

The **Bounce Mix** function, when offline mode is unchecked, will play back your session in real time while it prints the audio to a file. That is, if you selected 4 seconds of audio to export, then it will take 4 seconds to bounce, while if it's a 4-hour-long project, it will take 4 hours. The **Offline** checkbox allows for faster than real-time results. Depending on the number of tracks and plugins being used (and the complexity of the plugins), you can find your mix running many times faster than in real time – a number with an **x** symbol next to the progress bar denotes this.

Bouncing in offline mode might seem like the most practical option, no? Why should you have to sit there and wait, listening to a track you already mixed just to export it? It turns out that while great care has been taken to retain accuracy during offline bounces, there are still situations that arise where inaccuracies can occur. This is most prevalent with time-based plugins (such as reverbs/delay) and dynamics processing where attack and release times are calculated using a real-time clock. In most scenarios, the difference between an audio file processed in real time or offline will be negligible, but I have seen files fail quality assurance due to loudness errors, most likely introduced by a Limiter.

I also find that taking the time to listen to the entirety of a project in real time is both effective in evaluating whether all the mix and edit decisions made are being represented correctly and provides a cathartic moment to reflect on the completion of the project. That being said, there are certainly many situations where offline bounces are appropriate. With the majority of my podcast and spoken word work, the length of the projects, combined with the turnaround and expectations from the mix, make it a great opportunity to save time during the bouncing stage.

There are other methods for exporting a project as well, but using the **Bounce Mix** function gives some interesting options. Let's break those down and explain what each one does.

File Type

Right off the bat, you have the option to mix different file types. They are as follows:

- **WAV (BWF)**: This is known as Broadcast Wave Format and has become the de facto standard for almost all audio applications where file size is not a concern. Samples are encoded in **Pulse Code Modulation (PCM)**, which means there is no compression of the audio. The data encoded from a WAV file can be directly decoded by a digital-to-analog converter and is the most accurate of the outputted mix.

- **AIFF – Audio Interchange File Format**: Initially developed for Apple computers as a functional alternative to WAV files, AIFF tended to be more compatible when working in macOS due to how Apple CPUs operated. Currently, AIFF has no practical advantage, though there can be some legacy software that functions better with it.

- **MP3**: This is the lossy audio file format that compresses file sizes tenfold from WAV and disrupted the audio industry when it emerged with the popularization of the internet. MP3 uses psychoacoustics to strip "non-audible" data from a file, which in theory has little effect on what the listener experiences. In practice, using MP3 as a deliverable can remove much-needed information from a mix and should not be the first option you consider. However, it does still drive much of the podcasting and spoken word industries as its file size savings make it more practical for delivering audio to listeners.

- **MXF (OP-Atom)**: A digital media format that is designed to carry multiple types of media and metadata, including video and timecode. The OP-Atom version of it is stripped down, allowing only the most basic information to be contained, but MXF is the preferred file format for many digital distribution platforms.

- **MOV**: Exports the file to a QuickTime Movie. This option is only available when a video file is present and you wish to export the finished video file without having to send your audio to separate video editing software.

Mix Source

The **Mix Source** option allows you to select which output is being "recorded" to the file. This is beneficial when you are mixing different channel mappings or mixes from a single project and want to export them at the same time. For example, I could be working on a surround mix for a trailer to be shown in theatres, and for the same project, I could also be mixing a stereo version. Instead of having separate projects or using different moments in the timeline, I can set up specific busses in the track routing for those channel modes. This ensures congruency between the different versions.

If you're using Pro Tools Ultimate, you can perform multiple bounce sessions, or you can press the + button next to **Mix Source** to have multiple sources bounced simultaneously. When more than one **Mix Source** is selected and **File Format** is set to **Interleaved**, a new drop-down menu will appear below **File Format** called **Delivery Format**, which allows you to choose whether each mix is being exported as separate files or being mixed down to a single file.

Compression Type

For the most part, this menu will be inaccessible as the compression type is tied to **File Type**, but when you select a MOV file as the type, options for different audio compression codecs appear. The options available allow you to prioritize quality or file size for the export.

Add MP3

This option only appears when WAV or AIFF is selected. Checking this box will also export a separate MP3 file in addition to the uncompressed file. This is useful for situations where the final delivery needs to be in MP3 format, and you would like to keep a full-quality version for archival purposes.

File Format

There are three options available, depending on your chosen **File Type**:

- **Mono (summed)**: Mixes all the channels down into a single mono audio file. This is useful when you are exporting files with a single channel of audio, such as sound effects for game design, and for many spoken word applications where there is no stereo data.

- **Interleaved**: Multiple-channel audio will be exported to a single file with the different channels embedded into it. This is the most common way that audio can be exported as all the different channels are conveniently contained in a single file.

- **Multiple mono**: Every channel will be exported with the channel name appended to the filename. For example, a stereo track will be exported as two files, `BouncedAudio.L` and `BouncedAudio.R`. For MXF files and Digital Cinema Packages, Multiple Mono is required.

Bit Depth

There are typically three options here unless you are using MP3. Without going super deep into the topic of bit depth, the options usually align with their use cases:

- **16-bit**: Music, as this is the standard for CD audio

- **24-bit**: Motion picture, to give more headroom

- **32-bit float**: Virtually limitless headroom, allowing for greater flexibility concerning levels after the fact

While it may be tempting to export to a different bit depth than your project to try and "improve" the audio or save on data during export, the reality is that you may have more to lose than gain from this practice. Increasing the bit depth is not harmful (save for the extra data required), but if the original audio is recorded at a lower bit depth, Pro Tools can't make up that extra information to take advantage of it. Additionally, lowering the bit depth can introduce quantizing errors, where the output has noticeable artifacts unless proper processes, such as dithering, are applied. It is ultimately up to the client to decide what bit depth is the most appropriate, but it is best to keep this in mind before the project begins.

Sample Rate

There are a lot of options to choose from here, ranging from lowering the sample rate to 8 kHz to going all the way up to 192 kHz. Similar to bit depth, increasing the sample rate will not inherently improve the audio you are working on as this process is applied after the track is bounced, but it may be required to export to a different format for specific use cases. For example, it is possible to use audio samples in games made for older gaming consoles, but these typically need to be saved at a lower sample rate.

The **Pull Up** and **Pull Down** options are used to help to accommodate slight speed changes that can be introduced by different video formats. For example, an animator might work on a project that works at 30 **frames per second** (**FPS**) for convenience. However, most deliverables require the project to be at 29.976 FPS due to how video signals work. If they were to take your audio that was mixed with the 30 FPS video as a reference, it would be slightly faster and out of sync with the final project. I have first-hand experience with this and while it might seem like a small amount, it adds up over time. Instead of painstakingly trying to re-sync the material or make speed adjustments, you can use **Pull Up** to speed up the audio so that it matches the correct video standard.

Pad To Frame Boundary

A newer option in the latest Pro Tools, this is another benefit to those working with video. Pro Tools allows you to work at the sample level and export sample-accurate materials. For example, you can end a file at any point in the 48,000 samples per second if you're working at 48 KHz. Video is not as accurate. You can only start or stop a file when it lines up on the frame. So, if it's 24 FPS, that's as accurate as you can export. What this means when working with video is that an audio file should line up in that same interval, which in this example means that the file should end at an interval of 2,000 samples (48,000/24).

While not adhering to this interval shouldn't introduce significant issues, it's a good practice to ensure that things line up correctly for the video editor. With this option selected, Pro Tools will add a small amount of silence to the start and end points to make sure the file lines up properly in video editing software.

There's more...

I use **Bounce Mix** where appropriate, but often, I am printing directly to a track, as is detailed in the next recipe.

Printing directly to a track

Bouncing a mix is simple and has some nice export options, but it has one glaring issue – it's all or nothing. Whether you are amid a bounce or reviewing a file after the fact, if you notice a mistake or something you want to change, your only option is to re-export the entire thing. Even though offline bounces can take a long time if there are a significant number of plugins and tracks in place, wouldn't it be nice if there were a way to stop a bounce while it's printing to make a change, or just punch into a section to re-print just a spot after the fact? Luckily, there is! Printing to a track is the solution we'll use to achieve this result.

Getting ready

For this recipe, you'll need a Pro Tools session with several tracks of audio present. Make sure the **I/O** column in the track header is active. You can do this by using the quick select dropdown located at the top left of the tracks or going to **View | Edit Window Views** in the menu bar. Make sure that **Loop Record** is unchecked in the **Options** menu.

How to do it...

We'll create a Print track, route the audio to it, and record directly to it as the audio plays back. Follow along with these steps:

1. Select all the tracks in your session (click on the name of the first track, hold *Shift*, and click on the last track).

2. While holding *Shift + Option* (*Shift + Alt* on Windows), click on any track's output dropdown in the **I/O** column – it is the second drop-down menu.

3. Select **New track....**

4. In the **New Track** window, use the following settings:

   · **Format**: **Stereo**

   · **Type**: Audio Track

   · **Time Base**: Samples

   · **Name**: Print

5. Press the **Create** button.

6. In the track header for the newly created Print track, under the track name are four buttons – click the second from the left that is denoted by the letter **I**. This toggles **Input Monitoring** and is illuminated in green when it's active.

7. Begin playback by pressing the *spacebar* or clicking the play button in the transport. You should hear all the tracks as their audio is routed through the Print track.

8. Stop playback and click the first button on the left below the Print track's name – this is the **Record Arm** button; it's denoted by a circle and flashes red when it is active.

9. Press the *Return* key on your keyboard to return the timeline insertion to the beginning of the track.

10. In the transport in the toolbar, press the red circle button (**Record Arm**) to activate record mode.

11. Press the play button or press the spacebar and observe how the track is being recorded as it plays.

12. Stop the playback when the track has finished playing:

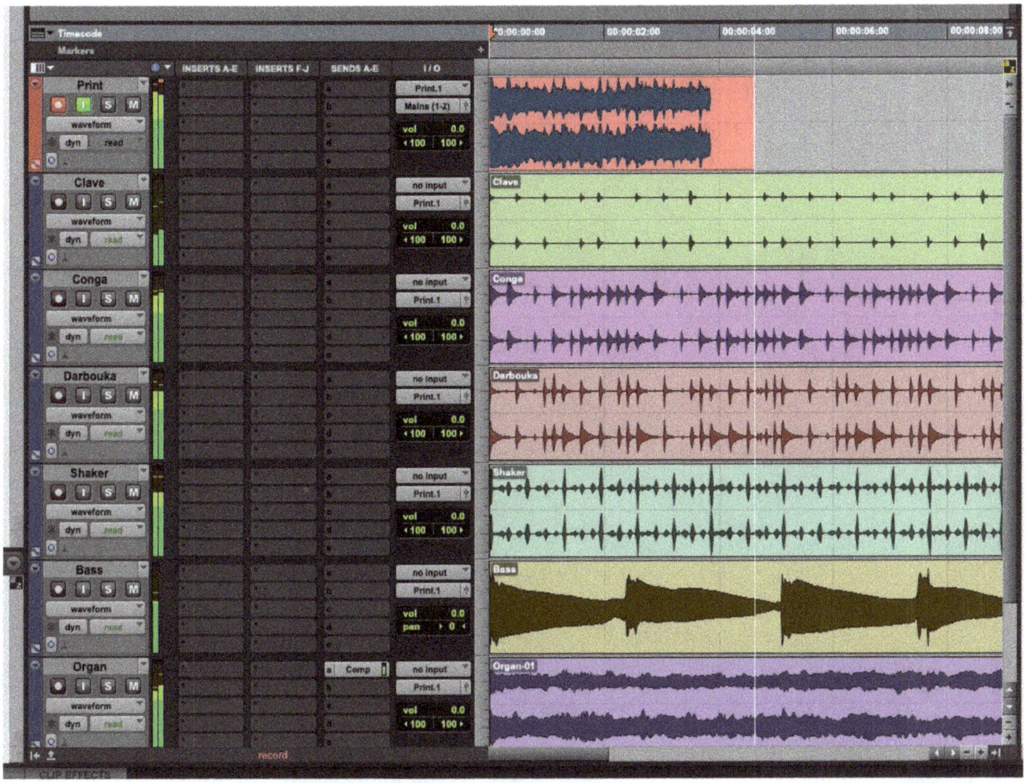

Figure 6.2: Printing to a track in action

13. Use **Grabber Tool** to select the newly printed track.

14. In the **Clip List** area (the sidebar on the right), click the triangle at the top right to bring down a drop-down menu (you can also right-click on the Print clip in the **Clip List** area).

15. Select **Export Clips as Files…**:

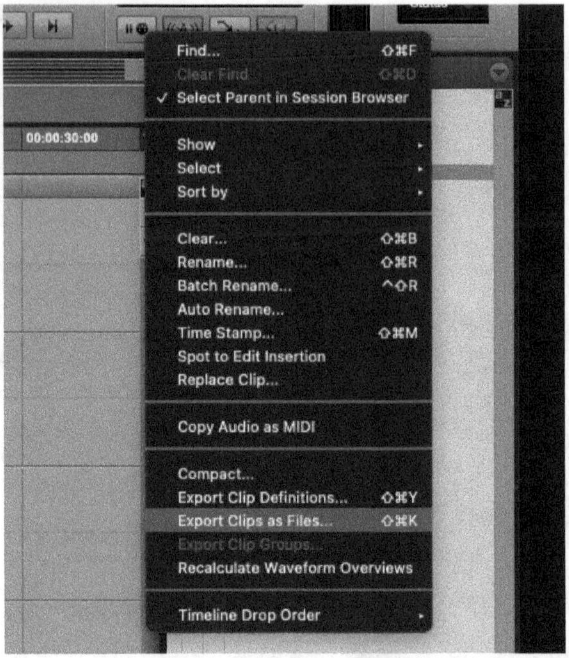

Figure 6.3: Exporting clips as files from the Clip List area

16. Chose the settings you want (see the previous recipe for details on the options available).

17. Click the **Choose…** button to select where on your computer these files will be saved – if you neglect to do this first, it will prompt you later.

18. Press **Export…**.

How it works…

Any audio track in Pro Tools can record the signal that is being routed to its input. Most Pro Tools users understand this concept when recording instruments, as the track input must be set to whatever hardware input you are using to connect the instrument or microphone. The beauty of Pro Tools is how audio that's routed internally can also be recorded. So long as a track has its output being sent to a record track, and that record track's input is being set to the same bus, then this will work. It is possible to set this up all manually using **Session I/O**, but I prefer the simplicity of using the **Output** dropdown to send the audio directly to a track.

When setting the outputs of multiple tracks, you can set them individually, but I prefer using the *Shift + Option* (*Shift + Alt* on Windows) modifier keys to apply the change to all selected tracks.

The **Input Monitoring** toggle ensures you can hear the tracks being played back when you are not recording them. If you want to hear the track as it is recorded, you can toggle **Input Monitoring** off. This is a good way to determine whether a change you're applying to a mix is the correct choice by comparing it to the already printed version.

There's more...

The exported file will retain the same name as the clip when using this method. You can always rename the file later, or if you double-click on the clip with **Grabber Tool**, the clip name prompt will appear, allowing you to change it beforehand. Clip names must be unique, so you may need to append a number to the clip name if it is already being used somewhere else. You can also change the name of the Print track as the clip that's been recorded will be named according to it.

Making changes to a Print track

One of the big advantages of printing directly to a track is the ability to make changes to segments of a print as opposed to bouncing the entire session again. We're going to go over how this is done with this recipe.

Getting ready

For this recipe, you will require a session with audio tracks routed to a print track that has already been printed. You can review the previous recipe to perform these steps. Make sure **Loop Record** is unchecked in the **Options** menu.

How to do it...

We're going to make a change to a mix by muting one of the existing tracks. Then, we'll print a portion of the mix and consolidate the printed clip so that it can be exported. Do this by following these steps:

1. Select one of the tracks in your session to be muted. Press the **M** button below its track name to mute it (the button will turn orange to indicate the mute is active).

2. Use **Select Tool** and highlight a short section of the session toward the middle of the track; a few seconds will do.

3. Set the Print track to **Record Arm** by pressing the record button below its name (the first button on the left, with a red circle).

4. Set the session to **Record** mode in the transport by pressing the red circle button.

5. Press play on the transport or press the spacebar to record. Since a portion of the track was selected, it should stop automatically:

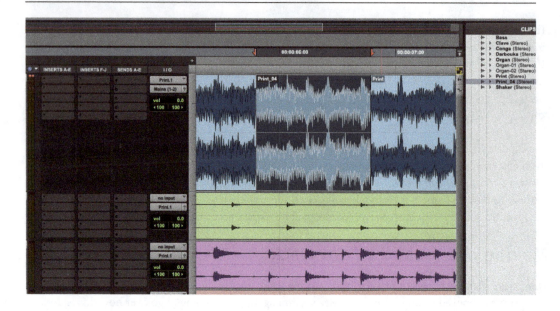

Figure 6.4: Recording over an existing Print track

6. You should now see a new clip that's recorded over the area that was highlighted in the print track. Use **Select Tool** and click and drag a selection so that it's a bit larger than the clip area.

7. Press *Command + F* (*Control + F* on Windows) to bring up the batch fades window.

8. In the middle of the batch fades window, set the crossfade options as follows:

    -   <<- **Link** ->>: `Equal Power`

    Crossfade Operation:

    -   **Create New Crossfades**: `Enabled`

    -   **Length**: `50 ms`

9. Press the **OK** button.

10. When you're prompted about invalid fade bounds, click **Adjust Bounds**:

Figure 6.5: Crossfades with adjusted bounds on the print track

11. Select the entire print track by pressing *Command + A* (*Control + A* on Windows).

12. In the menu bar, select **Edit | Consolidate Clip**.

13. In the **Clip List** area (the sidebar on the right), click the triangle at the top right to open a drop-down menu.

14. Select **Export Clips as Files…**.

15. Choose the settings you want (see the *Bouncing a project to a finished file* recipe for more details on the options available).

16. Click the **Choose…** button to select where on your computer these files will be saved – if you neglect to do this, it will prompt you later.

17. Press **Export…**.

How it works...

Functionally, making changes to an already printed track is the same procedure as printing the entire track, just focused on a specific area. The risk when making changes and printing in this fashion is that you may create an edit on a non-zero crossing in the waveform, which will result in a noticeable pop or click when it's played back. The best way to avoid this is to make the selection you re-print larger than the change you made. For example, if you make a change to a clip that's a few seconds long, then start your selection a second before that clip starts and end it a second after the clip ends. Theoretically, this means that the resulting wave will match up exactly and won't cause a pop. However, I still like to add a tiny crossfade just to ensure there is no issue.

You will get a warning about the crossfades being out of bounds. You can avoid this by extending the end point of the preceding print to overlap with the newly printed insertion, and similarly extending the start point of the next clip to an earlier spot. However, if this isn't accounted for, then these additions could overwrite your re-print.

The final step, consolidating the clip, merges all the existing clips and fades into a single cohesive clip, allowing you to export it in one shot from the clip menu.

Printing stems

Stems can be thought of as submixes for a project. In music scenarios, you can group your tracks into submixes, such as drums, guitars, vocals, and so on. In motion picture, these submixes are often broken down into dialog, music, ambiances, effects, and foley (though some of that can overlap). When you deliver a final mix, having the submixes printed as standalone stems provides some extra flexibility after the fact. For music, it's great to have stems should you wish to remix or remaster a song, or perhaps create a whole new version with different guest artists performing on it. Having stems means you can send those tracks along as opposed to the entire project, which is also useful if you consider all the plugins you may be using that aren't easily transportable.

In motion picture, stems have a specific purpose beyond the ability to remix. For foreign language productions where actors are hired locally to dub over the lines, the production will request a **Music and Effects** track (also known as an **M&E**). This is essentially a stem with everything except the dialog present. Not providing this can jeopardize a movie's ability to be distributed or cost the production team significant financial losses to hire another company to create these tracks.

Building upon the concepts in the previous two recipes, this time, we'll add intermediary tracks that will serve as the print tracks for the stems. We'll use Aux tracks this time to provide one more opportunity for inserting effects before a track is printed.

Getting ready

For this recipe, you will need a Pro Tools session with multiple tracks of audio (at least nine). A music session with different instruments is a good example. Make sure the **I/O** column is visible in the **Track** header. You can do this by using the quick select menu at the top left of the track listing, or by going to the menu bar and selecting **View | Edit Window Views**.

How to do it...

We're going to route groupings of tracks to Aux tracks, set the output of each Aux to a print track, then send the outputs of those print tracks to another **Master Print** track. When recording, this will print all the stems and the master mix simultaneously. Follow along with these steps:

1. Select the first grouping of tracks by clicking on the name of the first track, holding *Shift*, and clicking on the last track you wish to be grouped in a stem – if you're using an existing session, this could be instrument groupings, or you can simply use the first three tracks.

2. In the **I/O** column for one of these tracks, click the **Output** dropdown (the second from the top) and select **New track....**

3. In the **New Track** window, use the following settings:

   - **Format: Stereo**
   - **Type**: Aux Track
   - **Time Base**: Samples
   - **Name**: Stem 1 (feel free to change this to something more appropriate)

4. Repeat *steps 1* to *3* for the other tracks in your session, creating two more stems called Stem 2 and Stem 3.

> **Note**
>
> The next two steps are optional, but I find them much easier to keep a workflow organized.

5. Click on the **Stem 1** name to select it, hold *Command* (*Control* on Windows), and click on **Stem 2** and **Stem 3**.

6. Click and drag **Stem 2** to the top or bottom of the track list to group all three.

> **Note**
>
> Where you place your stem's Auxes and subsequent print tracks is a matter of personal preference. When mixing, I conceptualize the audio "bubbling up" toward the top of the tracklist. Many mixers I work with also like to think of the track audio "trickling down" toward the bottom. Whether or not you use a physical mixing controller also tends to have an impact, as the top tracks will map to the left of the mixing console, and the bottom tracks will be on the right.

7. Hold *Command* (*Control* on Windows) and click on the third button from the left (**Solo**) under each stem Aux track – the **S** button will be grayed out, indicating that these tracks are solo-safe:

Figure 6.6: A session with stem Aux tracks set

8. On the **Stem 1** Aux, click on the **Output** dropdown in the **I/O** column and select **New track…**.

9. This time, use these settings to create a stem print track:

    ·   **Format: Stereo**

    ·   **Type: Audio Track**

    ·   **Time Base: Samples**

    ·   **Name: Stem 1 - Print**

10. Repeat *steps 7* to *8* for **Stem 2** and **Stem 3**, creating audio tracks named `Stem 2 - Print` and `Stem 3 - Print`.

    These new stem Print tracks will be created next to the existing Aux tracks. Once again, you can move them to the top or bottom if you want easier organization:

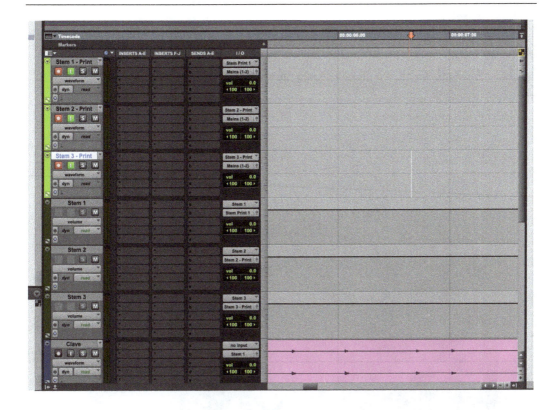

Figure 6.7: Sending Aux tracks to Print tracks

11. Click on **Stem 1 - Print** to select it.

12. Hold *Command* (*Control* on Windows) and click on **Stem 2 – Print** and **Stem 3 – Print** to add them to the selection.

13. Hold *Option + Shift* (*Alt + Shift* on Windows) and, under the stem Print track names, click the two left buttons to activate **Record Arm** and **Input Monitoring**.

14. While still holding *Option + Shift* (*Alt + Shift* on Windows), select the output dropdown under the **I/O** column and select **New track…**.

15. Use the same settings from *step 3* to create an Aux track, naming it Summing Aux.

16. Hold *Command* (*Control* on Windows) and click the **Solo** button (**S** under the track name) on the Master Aux to make it solo-safe.

17. In the **Summing Aux** track header, click the **Output** dropdown in the **I/O** column and select **New track…**.

18. Follow the settings for *step 9* and create another Print track, naming it Master Print.

19. Click the two left buttons under the track name to enable **Record Arm** and **Input Monitoring**.

20. In the transport, click the **Record Enable** button (red circle).

21. Press *Return* (*Enter* on Windows) to bring the play head to the start of the track.

22. Press the spacebar or the play button in the transport and watch the stems and master print track print simultaneously:

Figure 6.8: Stems and Master printing at the same time

How it works...

Both aux tracks and audio tracks allow you to route audio to both their input and output, so they functionally do the same thing, with the bonus of an audio track being able to record audio to it. With this in mind, any routing setup you create for submixes with Aux tracks can also function as a way to print stems. It's possible to not use Aux tracks at all and send audio directly to print tracks, but as we'll learn later in this chapter, Aux tracks can be leveraged to place inserts on the audio that's routed to it before it's printed.

This method can also be used to print specific submixes, such as Music and Effect tracks, and using Sends to split the signal (see using the *Splitting audio with sends* recipe in *Chapter 4* for details).

There's more....

Just like in the previous recipes, you can select all the print tracks with **Grabber Tool** and export the clips as files in the **Clip List** area. You can also punch in and record over sections of the tracks if you make changes.

You can also disable **Input Monitoring** on a stem to use it for edits/changes, which can sometimes make things easier.

Understanding LUFS

LUFS (also known as LKFS) is an algorithm that calculates the way loudness is measured for modern audio programs. The first iteration of this standard was introduced in 2006, with the current revision being released in 2015. Before LUFS, other standards relied on a combination of signal playback and sound level pressure meters to ensure that the audio was set to the correct level. The ability to measure an absolute loudness level with the signal alone allows for much more accurate and translatable mixes to be delivered and opens up the ability for delivery to be performed with much more accessible tools.

The beauty of using LUFS as a target for delivery when mixing your work is that you can mix to a loudness level you are comfortable with. While there are standards that dictate what levels are supposed to be ideal, the reality of the space you're mixing in or the tools you use can make it not practical to aim for theater-level loudness. Additionally, if you are mixing for long periods at high levels, you can fatigue your ears and inadvertently compensate by making your mixes louder as time passes. Using LUFS to measure the signal's loudness means you can set your monitors or earphones to the loudness you feel is right for you to mix with, and know the delivery level will be appropriate.

While Pro Tools does not have a built-in LUFS meter, the Avid Pro Limiter does include one. This is part of Pro Tools Studio or higher. For this recipe, however, we will be using a third-party plugin, YouLean Loudness Meter. This excellent plugin has a great set of visual tools that make it easier to understand how LUFS is calculated, and what's important to consider when aiming for a loudness standard. What's more, the free version is perfectly useful for most applications, with the Pro features being nice-to-have touches that increase the productivity workflow in professional environments. At about $50, the Pro version is also not overly expensive if those features meet your needs.

Getting ready

You'll need to install YouLean Loudness Meter for this recipe, so visit `https://youlean.co/youlean-loudness-meter/` and download the installer there. Pro Tools will need to be relaunched if you already have it open so that the plugin will appear.

You'll also need a session with multiple audio tracks routed to an Aux track. The previous recipe showed you how to set one up, though you can set up a session with the track's output being routed to an Aux track named `Summing Aux`.

In your Pro Tools session, make sure that the **Inserts** column is shown in the **Edit Window** area. You can do this by going to the menu bar and selecting **View | Edit Window Views** and making sure the appropriate items are checked, or by using the **Edit Window View selector** dropdown directly above the track headers on the left side of the **Edit Window** area.

How to do it...

We'll add YouLean Loudness Meter and a Summing Aux insert to measure the loudness of a session before its signal is sent to a print track. Follow along with these steps:

1. In the **Summing Aux** track, click on the first available **Insert** slot and select **multichannel plug-ins | Sound Field | YouLean Loudness Meter 2 (stereo)**:

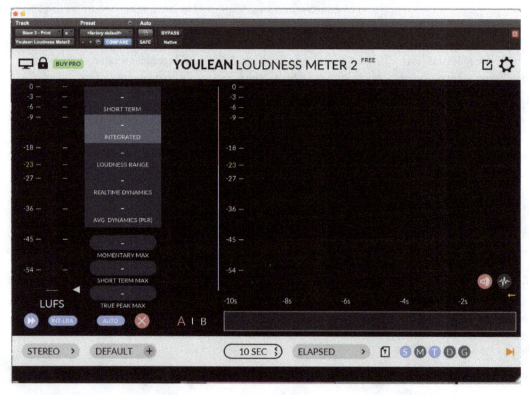

Figure 6.9: YouLean Loudness Meter

2. Press the spacebar or press play in the transport to start the playback.

3. Observe the two areas labeled **Short Term** and **Integrated** as your audio plays. **Short Term** should change quickly, while **Integrated** should stabilize to a steady value over time.

4. Observe the value for **True Peak Max** toward the bottom middle. If this is a positive amount, it should be in red.

5. Observe the meter on the left. By default, the value of **-23** should be highlighted yellow, with levels below it in white and levels above it in red.

6. Toward the bottom left, click on the button that says **Default**.

7. From the list, select **ASWG-R001 PORTABLE**.

8. Notice that the yellow target level on the left has changed to **-18**:

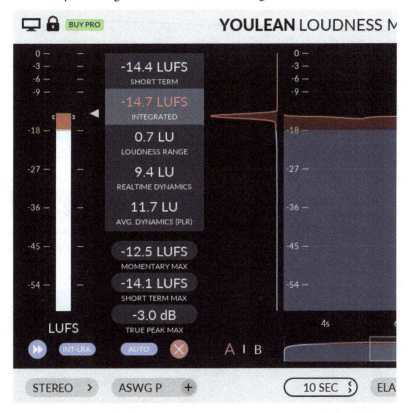

Figure 6.10: YouLean preset to ASWG-R001 Portable (-18 dB LUFS)

9. Adjust the volume levels of other tracks up and down and observe the behavior on both the **Short Term** and **Integrated** levels, as well as the graph to the right of the plugin window.

10. Stop the playback and use the navigator at the bottom right to move left and right on the graph and see the changes in loudness over time.

How it works...

LUFS is measured with two methods, **Short Term** and **Integrated**. **Short Term** refers to what the loudness is at that moment in the program, whereas **Integrated** is the overall loudness of the entire program from start to finish. The loudness measurements use a digital meter, with **0 dB** being the absolute loudest threshold and everything quieter being measured in the negative below it. Different standards have been declared over time, which is what the different presets under the **Default** button refer to. If you scroll down further, you will see presets for various online streaming services such as **Spotify**, **YouTube**, and **Netflix**, but these features are locked behind the Pro (paid) version.

The standards these different services require are, for the most, part publicly available or are provided to the production team by the service as a Quality Assurance stipulation. While the presets in YouLean are convenient, if you know what you are aiming for, you can use the meters and values listed and adjust accordingly. If you go over the loudness required by a service, one of two things can happen:

- The project will be rejected due to not complying with Quality Assurance
- The project will be subjected to a "loudness penalty," where the streaming service will lower the volume to meet its targets

Most services will use the **Integrated** loudness to determine whether your audio falls in spec, so while it's possible to simply increase or lower the overall volume of a track before it's printed to meet the spec, the purpose of the **Short Term** loudness and the graph is to identify where in the mix you may be going over. Focusing on those moments and lowering them accordingly is a more practical way to follow quality control targets while still retaining the overall decisions you've made in your mix.

There's more...

You can see the loudness penalty in action on any YouTube video in a desktop browser. Right-click on the video and select **Stats for nerds**. At the top left of the video, you will see several details, one of which is **Volume/Normalized**. If the audio of the stream is on target or under (-14 dB LUFS for YouTube), it will show **100%/100%** and the difference from the target in brackets (for example, content loudness -1.3 dB).

If the content was mixed too loudly, it will show how much it was lowered under the **Normalized** denominator (for example, 100%/62% content loudness 4.1 dB).

Using Limiters for quality assurance and compliance

We've seen what LUFS are and how they are used, and gone over how to adjust the volume for a project accordingly to meet loudness standards. While it's good to be able to focus on trouble spots, you have tools at your disposal to make mixing for specific targets easier. Mainly, Limiters are your best friend in this area. Limiters can work as a "brick wall" that prevents anything from going over a specific peak level. They can also be used as a "maximizer," making things louder to meet a target, depending on how they are used.

I personally always like to mix somewhere around -24 dB LUFS to give me lots of headroom, regardless of the medium I'm mixing in. Many delivery formats are close to that, and I may adjust accordingly at the mix stage to ensure I hit those targets, but many, many online streaming platforms target much louder than that. Instead of losing that headroom while I mix, I opt to use a Limiter's "maximizer" capability to increase the overall loudness of a project without it having a massive effect on the tone or overall balance of the mix during the print phase. This does affect dynamic range, so this method is best left for projects where having strong dynamics is not pivotal to it.

Getting ready

You'll need to install YouLean Loudness Meter for this recipe, so visit `https://youlean.co/youlean-loudness-meter/` and download the installer there. Pro Tools will need to be relaunched if you already have it open to have the plugin appear.

You'll also need a session with multiple audio tracks routed to an Aux track. The previous recipe showed you how to set one up, but you can set up a session with the track's output being routed to an Aux track named `Summing Aux`.

In your Pro Tools session, make sure that the **Inserts** column is shown in the **Edit Window** area. You can do this by going to the menu bar and selecting **View | Edit Window Views** and making sure the appropriate items are checked, or by using the **Edit Window View selector** dropdown directly above the track headers on the left-hand side of the **Edit Window** area.

How to do it...

For this recipe, we'll be adding a Limiter to a Summing Aux's Insert, followed by YouLean Loudness Meter to see how the Limiter's settings affect the overall loudness. Follow along with these steps:

1. In the Summing Aux's **Insert** column, click on the first **Insert** slot and select **multichannel plug-in | Dynamics | Maxim**.

2. At the top right of the **Maxim** plugin window, click the red box (**Target Button**) to keep the plugin window when another plugin is activated:

Figure 6.11: The Maxim plugin window

3. In the Summing Aux's **Insert** column, click on the second **Insert** slot and select **multichannel plug-in** | **Sound Field** | **YouLean Loudness Meter 2 (stereo)**.

4. Start the playback with the spacebar or the play button in the transport and observe what YouLean Loudness Meter is measuring for LUFS (both **Short Term** and **Integrated**).

5. In the **Maxim** plugin window, click the **LINK** button in the bottom middle of the window to activate it.

6. Click and slowly drag the **THRESHOLD** slider down toward -20.0 dB and observe the effect it has on the **LUFS** readings in the YouLean meter plugin:

Figure 6.12: Maxim set to -20 dB – linked

7. Click the **LINK** button to deactivate it.

Loudness warning

For this next step, things are about to get LOUD. Take precautions with your monitoring devices, whether you're using headphones or not.

---

8. Click and slowly drag the **CEILING** slider up and notice the effect on the **LUFS** meters in the YouLean plugin – stop before it gets too loud.

9. Click and slowly raise the **THRESHOLD** slider back up and notice how it affects the **LUFS** meter in the YouLean plugin:

Figure 6.13: Maxim set to -12 dB Threshold, -0.01 dB ceiling

How it works...

Limiters, in concept, work very much like Compressors (see *Chapter 5*, recipe *Using compression to control dynamics*, if you are not familiar). You can use any compressor that allows for a fast attack and release, and a high ratio to effectively work as a limiter. However, in practice, most brick-wall limiters and maximizers are designed with a different end goal in mind – that is, controlling loudness as opposed to dynamics. Therefore, the options given to the user for manipulating how they work are very different.

You are essentially given two sliders to work with. If the specifications you are provided with include a max peak, then it's best to start there, with the **CEILING** slider set to that level. This will prevent any signal from going higher than what you've set. The next consideration is **THRESHOLD**, which will set at what point the limiter engages. The combination of the two settings will determine the resulting output. If the threshold and ceiling are linked, or the ceiling is lower than the threshold, then the result will be quieter and have a lower LUFS reading. If the threshold is set lower than the ceiling, then the Limiter will increase the loudness of the signal and you will have a higher LUFS.

This is an easy way to hit a target LUFS if you choose to mix at a lower level for more headroom as I do. I mix at around **-24 dB LUFS**, then apply a limiter with a threshold set to around **-10 dB** to bring it up to **-14 dB LUFS**, for example.

There's more...

Maxim is an okay Limiter, in that it provides some nice visual cues and mimics the functionality of most brick-wall limiters, but it is prone to distortion/clipping and will often leave you with a muddy mix in the end compared to most other Limiters on the market. While you have plenty to choose from, you'll hear many audio professionals endorse Waves L3 and FabFilter Pro-L2. The Avid Pro Limiter is also a decent choice if you have a higher tier of Pro Tools that offers it. Both the Avid Pro Limiter and FabFilter's offerings also offer a LUFS meter built in, giving you an all-in-one package, but if cost is a concern, Waves L3 is a long-standing tool that I use on almost every project.

As with many plugins, different Limiters not only offer different benefits and drawbacks in terms of how they operate but they can also provide different tonal characteristics. Your best bet is to download a trial and see whether you like the results.

Consolidating and archiving a project

We like to hope that once a project is done, that's it – we can call it a day and move on to another project. In truth, you may need to revisit a project long after it has been released. I once had a project I worked on that needed a change in music due to some unfortunate licensing issues 10 years after it was completed. Had I not held onto an archive, I could have still fallen back to the stems, but the way the music was mixed in this specific project would have made it a lot harder.

Now, it's possible to simply compress or just copy an entire session folder as is to an external storage system, but often, a project's clip list gets bloated with material that's not needed for long-term archiving. It's also possible that you may have (intentionally or not) audio files that are saved outside of the session folder. We'll use the features in the **Clip List** area to remove unused files, then the **Save Project As...** dialog to consolidate all the media used in a project to a separate location to be offloaded from your computer.

Getting ready

For this recipe, you will need a Pro Tools session with some audio tracks with clips in them.

How to do it...

First, we're going to remove unused files from the session, then save a copy to consolidate the files. Follow along with these steps:

1. Save your session with *Command + S* (*Control + S* on Windows) or go to **File | Save**.
2. If you have any inactive tracks, you can delete them to save space – *Command* and click (*Control* and click on Windows) on the tracks you wish to delete, then right-click on one, and choose **Delete....** Choose **Delete** when prompted.

3.   In the **Clip List** area, click the triangle at the top right and choose **Select | Unused**:

Figure 6.14: The Select | Unused option in the Clip List area

4.   In the same menu, select **Clear…**:

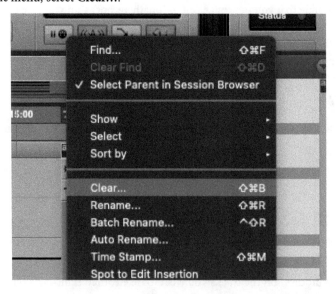

Figure 6.15: The Clear action in the Clip List area

5. When prompted with options to **Remote** or **Delete**, click **Remove**.

6. In the menu bar, select **File | Save Copy In…**:

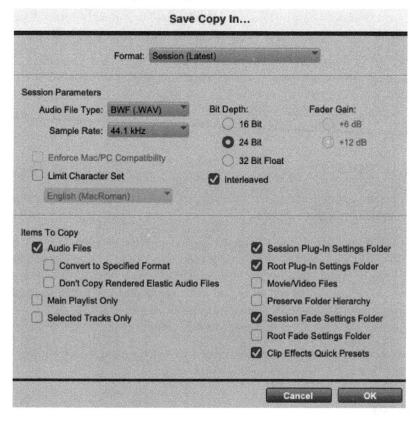

Figure 6.16: The Save Copy In… dialog

7. Under **Format**, chose **Session (Latest)**.

8. For **Session Parameters**, choose the options that match your project's current settings.

9. In the **Items To Copy** section, make sure you have **Audio Files** checked.

10. If you wish to have other items, such as **Movie/Video Files**, included, you can go ahead and check those too.

11. If you have been using playlists, and only want the main playlists to be exported, check **Main Playlist Only**.

12. Click **OK**.

13. Choose a name and location for the exported Pro Tools session and click **Save** – this will create a folder containing the session file, as well as audio files and other pertinent folders inside it.

How it works...

Using both the **Clip List** area's **Select Unused** function and its **Remove** function allows you to get rid of files from the session that are no longer being referenced in and of the tracks. If you wish to retain these clips for whatever reason, you may want to consider placing them in an inactive track.

Using the **Save Copy In...** command with **Audio Files** checked will take all the audio files within a project and save them to a new session. Since the unused files were removed, you will be left with just the files that are in use, and potentially save on file space. This will also consolidate audio files from areas that are outside of your session, in case you had imported audio from other folders and chose not to copy them.

Creating templates

It might seem counterintuitive to make a template after you've completed a project, but this is the best time to do it in my opinion. Let me explain.

It's possible to forecast some of the needs of a project, but as the project progresses and you get further down the workflow, you will inevitably make changes and additions to how things were set up from the beginning. Some of these changes might be very specific to the project you are working on, but many things, such as audio routing and setting up sends and print paths, will be foundational. It's these changes you want to capture and set up as templates moving forward, especially if you work on a series of regularly released programs that has consistent needs.

Let's go over a few of the options we have when creating templates, along with their pros and cons. After, we will discuss some strategies for making better templates.

Getting ready

For this recipe, you will need a Pro Tools session with some audio tracks that have audio clips in them.

How to do it...

We're going to save a session as a template and make it available in the Pro Tools dashboard when it opens. Follow along with these steps:

1. In the menu bar, go to **File | Save As Template...**.
2. In the **Save Session Template** menu, select **Install template in system**.
3. In the **Category** dropdown, select **Add Category...**.
4. In the **New Category** window, enter Packt (we'll remove this later).
5. In the **Name** field, enter Template Example.

6.   Make sure **Include Media** is unchecked:

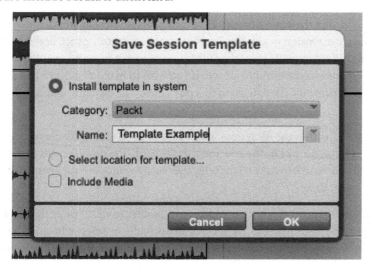

Figure 6.17: The Save Session Template window

7.   Click **OK**.

8.   In the menu bar, select **File | Create New…**.

9.   In the dashboard, check the **Create From Template** option.

10.   Select **Packt** from the **Template Group** dropdown.

11.   Click **Create**.

12.   Observe how all your previously completed tracks have been retained, but the audio files from the project are not present.

13.   Now, let's remove this example template. Go to the menu bar and select **File | Save As Template…**.

14.   In the **Category** menu, select **Reveal Session Templates** (this should be in your **Documents** folder under **Pro Tools | Session Templates**).

15.   Select the **Packt** folder and move it to the trash.

16.   Back in Pro Tools, in the menu bar, select **File | Create New…**.

17.   Click on the **Template Group** dropdown and notice that the Packt option is no longer listed.

How it works...

Pro Tools template files are the same thing as a Pro Tools session file but with a different extension. You can change the file extension of any Pro Tools file from `.ptx` to `.ptxt` and it will behave like a template. The only difference with a template file is that it cannot be saved over itself. It will always require a new session to be created when opened. If you wish to make changes to a template once it's been created, simply change the file extension to `.ptx`, make the changes, and save it, then change it back to `.ptxt`.

All templates are stored within the filesystem, so it's easy to go in and make changes as you wish to have them appear as you like within the Pro Tools dashboard. The dashboard is just a glorified file explorer, but it's still convenient to have all the templates ready within the app than trying to navigate to a specific folder every time.

If you wish to send a template to another computer, or to another engineer to work with, then you can either copy a template file from your Pro Tools **Documents** folder or **Select a location for template...** in the **Save Session Template** options menu. We'll get into what including media means in a bit.

There's more...

Now that the mechanics of making templates are out of the way, let's talk about what makes a good template, or at least the things I like to include in mine.

Depending on the project, I like to have all my Aux and Print tracks set up with the correct routing applied. I also like to have all my tracks named according to what audio will be going in them (if you've been using Audio 1 and Audio 2, this is a very good time to stop doing that). For certain tracks, I like to duplicate a few for extra "buffer;" it's easier to delete them after for me – something such as effect tracks for an audio drama. I might only need three or four, but I'll put eight in the template for that extra leeway.

Here are some other things I like to put in place before I save a project as a template:

- Track colors
- Track folders
- Groups
- Plugins and effects

Another thing I like to put in place is an inactive track at the bottom of the session with the name - - - to serve as a spacer. You can place spacer tracks like this anywhere, but when using mixed templates and sessions where I expect to import material from an OMF/AAF or another session, I like to have a separator below them, just to keep things clear for me when I'm copying media to specific tracks:

Figure 6.18: An example mix template for one of my ongoing projects

7

Considerations
for Music Production

The tools and workflows for music production can vary depending on the genre of music being created, and how that music is being recorded. A bluegrass band may simply need one mic in a room as the players move around it (I've seen this in person a few times – it's quite impressive!), whereas an electronic music producer may have no microphone recordings at all and either generate the sound within Pro Tools using Instrument tracks or externally with different synthesizers and record the audio into Pro Tools. Given the massive variety of ways a music project can be implemented, this chapter cannot encompass all methods. Instead, the goal is to provide a number of workflows and recipes that can be applied to different situations.

With this in mind, here are the topics we'll be covering:

- Setting the correct mix level
- Keeping drums in phase
- Using expanders to create more "punch"
- Applying sidechain compression
- Working with vocals
- Manipulating tempos

Technical requirements

The recipes in this chapter require at least Pro Tools Studio.

Example files for each recipe can be found at `https://github.com/PacktPublishing/The-Pro-Tools-2023-Post-Audio-Cookbook`.

Setting the correct mix level

One of the most common conversations among both beginners and seasoned music engineers is about the right loudness level for mixing music. There is some personal preference at play but the most important thing to consider is **Headroom**, which is how much dynamic range a mix has *over* the target mix level. For instance, if I am mixing a piece of music to a target of **-10 dB LUFS**, that gives very little room above that target for signals to go over before they distort. However, if I set my target to **-24 dB LUFS** I have a lot more range to play with.

This might seem like something that is not consequential – can't you just compensate by turning certain things down? The reality is our ears and brains don't perceive loudness equally across the spectrum of frequencies. You can look at something called the **Equal Loudness Curve** if you want to get the exact details, but the core concept is that we as humans have evolved to hear sounds that fall within the range of human speech better than others. What does this mean for your mix? Instruments and sounds in the lower ranges especially will require a lot more power and signal strength to be perceived at the same loudness as sounds in the higher frequencies.

If you set up your mix environment correctly, then you will have plenty of headroom to push those instruments to the power they need compared to the sounds that don't need as much. When a beginner engineer is presenting a mix to me that feels flat or lacks low end, it's almost guaranteed that they were mixing at too loud of a target level.

With this recipe, we'll look at how to set the loudness of your mix to a target that will provide adequate headroom and dynamic range to get a good mix.

Getting ready

For this recipe, you will need a Pro Tools session with a stereo audio track and YouLean Loudness Meter installed (see *Chapter 6, recipe Understanding LUFS*). You'll also need a piece of music that is aligned with the style or genre of what you are aiming to achieve with your mix. This doesn't need to be a song you made yourself, just one you think is well produced and sounds good – we'll be using this as a reference.

Make sure in your Pro Tools session that the **Inserts** column is shown in the **Edit** window. You can do this by going to the menu bar and selecting **View | Edit Window Views** and making sure the appropriate items are checked, or by using the **Edit Window View Selector** dropdown directly above the track headers on the left side of the **Edit** window.

How to do it...

We're going to place a piece of reference music in our session, adjust its gain to have it measure **-24 dB LUFS**, then adjust the loudness of our listening system to make it sound comfortable for the mix. Follow along with these steps:

1. Place the reference audio clip in an audio track.

2. In the **Inserts** column in the track header, click on the first insert slot and select **multi-mono plug-in | Other | Trim**.

3. Click the red square button in the top right of the plugin to have it stay onscreen.

Figure 7.1: The Trim plugin

4. Click on the next available insert slot and select **multichannel plugin | Sound Field | YouLean Loudness Meter 2**.

5. Start playback on the transport or press the *spacebar*.

6. In **YouLean Loudness Meter**, observe the value for **Integrated** (this may take some time to gather an accurate reading if the piece of music you selected has a large dynamic range).

7. In the **Trim** plugin, adjust the gain level until the loudness reads -24 dB LUFS.

Dialing in the gain

You will most likely need to decrease the gain to achieve the -24 dB LUFS reading. The amount should be proportional to the **Integrated** reading. That is to say, if the initial level is -14 dB LUFS, you'll need to set the **Trim** plugin's gain to -10 dB to bring it down to -24 dB. You may need to stop and start playback to get a proper reading.

---

8. Adjust the loudness on your hardware until you find it at a comfortable level.

> **Adjusting the hardware level**
>
> Your hardware could be the system audio if you are not using an audio interface, or it could be the speakers if the interface you are using doesn't have a volume adjustment (rare, but it does happen). The key thing here is you are not using Pro Tools itself to change the loudness of your monitoring environment after you've set the correct level for the reference track. What is a comfortable level can be highly subjective, but your goal is to set it at something that is loud enough to hear all the details, while not fatiguing your ears.

Your loudness is now set. You may want to place a mark of some sort on your hardware (I use a label and marker) to be able to easily return to this level later.

How it works...

As mentioned earlier, to get the best mixes you need headroom for your mix to be able to push the signal properly. An analogy I like to use is photo printing in the digital space. If you don't have a properly calibrated display, when you go to print an image, it will look incorrect. If your display was too dark, then you will have increased the brightness to compensate or vice versa. With music, you need to set the loudness of the hardware you are using to give you the best results. This can be very subjective, which is why I recommend using an existing song that best fits the same goals you are aiming for in your mix. If you are trying to mix an electronic music piece, you wouldn't use a heavy metal song as a reference.

When you turn the gain down on the track, you typically need to increase the volume output of your hardware. This means the nominal level your track is being mixed in will be much lower than a conventionally mastered and released track, giving you that headroom you need. Once the mix is done, then loudness can be applied in the mastering stage (more on that later in this chapter).

Remember that once you've set the levels for your mix, you do not want to touch the hardware levels unless you can easily return to them. I see this often where a novice mix engineer will be presented with something that sounds too loud, so they reach for the volume knob on their hardware. Avoid this, and instead, turn the clip gain or volume down on that track.

There's more...

There are tools that can be used to calibrate a listening environment, and these are expected in a professional space. A **Sound Level Meter** is a device that measures the fluctuations in air pressure at the point it's being directed to. The typical process is to generate **Pink Noise at -20 dB** in a track (this can be done as an **Insert** plugin or with AudioSuite using **Signal Generator**), then adjust the monitor levels until they read 85 dB on the meter. This level can also be subjective, though, with some engineers preferring to mix at a lower level (most don't aim for higher to avoid ear fatigue). This method also doesn't really work with headphones, which is why I like using a reference track as an alternative.

Keeping drums in phase

Recorded drums can be one of the most challenging aspects of a mix. Getting the punch and power out of a drum kit while making sure it sits well in the mix can be difficult, and the first order of business is to check the phase of tracks. When audio signals are mixed, if the peaks and valleys of the waveforms are lined up, it will reinforce the sound. If the signals are not aligned, you instead will get a cancelation of signals, which can remove sounds entirely or make things sound strange, almost as if you are listening to them through a pipe.

Even when a drum kit is set up with mics situated super close to the instruments and gating or other noise reduction is used, there is still some bleed that occurs between the tracks, and this can cause phase cancelation issues. Luckily, it's pretty easy to check phase and we're going to do that in this recipe.

Getting ready

For this recipe, you will need a Pro Tools session with a drum kit recording with separate mics and tracks for each piece of the kit. This can be difficult to obtain, but we are fortunate to have a full kit recording provided by Christine Awasthy. This can be obtained in the GitHub project linked in the *Technical requirements* section at the beginning of this chapter.

Make sure, in your Pro Tools session, that the **Inserts** column is shown in the **Edit** window. You can do this by going to the menu bar and selecting **View | Edit Window Views** and making sure the appropriate items are checked, or by using the **Edit Window View Selector** dropdown directly above the track headers on the left side of the **Edit** window.

How to do it...

We're going to go track by track and align the phase of the drum kit to get the best sound out of the tracks. Follow along with these steps:

1. If you haven't already done so, import the audio tracks for the drum kit into your session – the order of the tracks isn't super important, but I like to lay them out in the order that I will be checking phase:

   - Snare (Top)

   - Snare (Bottom)

   - Kick

   - HiHat

   - Overhead Left (OH L)

   - Overhead Right (OH R)

   - Tom 1

- Tom 2

- Floor Tom

2. Hold *option* (*alt* on Windows) and click on the first available **Insert** slot, then select **plugin |
   Other | Trim**.

3. In the Pro Tools menu bar, select **Options | Solo Mode | Latch**.

Solo modes

There are two main solo modes that are commonly used in Pro Tools, **Latch** and **X-OR**. As
indicated in the menu, X-OR will cancel the previous solo. For most of the work I do, I prefer
X-OR mode, as it allows me to quickly solo between tracks for comparisons without having
to undo the previous solo. If you prefer this mode, then you can use the *shift* key to solo
additional tracks.

---

4. Begin playback with the transport or by pressing the *spacebar*.

5. Solo both **Snare** tracks by clicking the **S** button beneath their track names.

6. In the **Snare (Bottom)** track's header, click the **Trim** insert to bring it forward.

7. Towards the right side of the plugin window, you will see a button that looks like a circle with a
   line drawn through it (ø) – this is the **Invert Phase** button, click it to enable it and you should
   hear a noticeable improvement in the sound of the snare drum.

Figure 7.2: The Invert Phase button

8. Solo **Kick Drum**.

9. Click on the **Trim** insert in the **Kick Drum** track header.

10. Click the **Invert Phase** button and listen to see whether the sound has improved. If it hasn't, disable it – you may need to toggle it a few times to compare the difference.

11. Continue down the tracks, toggling the phase in the **Trim** insert and comparing whether the tracks are improved or not, taking care that the Overhead tracks' phase setting matches.

12. Hold *option* (*alt* on Windows) and click on any solo button to remove them all.

13. Adjust the panning on each track to your liking and enjoy the sound of a properly phased aligned mix.

How it works...

The easiest way to see how phase cancelation works is to examine the **Snare (Top)** and **Snare (Bottom)** tracks. If you zoom in enough, you can see that the shape of the waveforms is almost exactly opposite (see *Figure 7.3*). This is due to how the mics are positioned and spaced relative to the snare. You will find this is often the case with snare drums, as it's common practice to use two mics – one on top to capture the impact on the drum skin, and one on the bottom to capture the sound of the snare brushes.

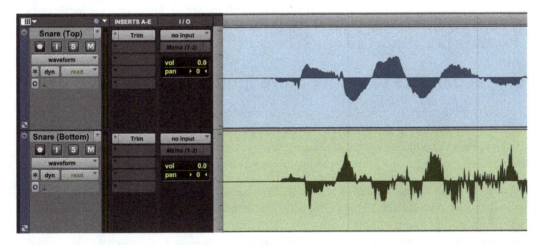

Figure 7.3: Snare tracks with inverted phases

In order to correct this, the signal's phase must be inverted. The peaks in the signal will become valleys and vice versa. If you want to see this visually, you can select the **Snare (Bottom)** clip and use **AudioSuite | Other | Invert** to see the result. Now the signals will align. Using an invert phase button in a plugin achieves the same effect.

With each additional instrument added to the mix, parts of the previous kit can either add or subtract from the overall sound according to its phase. While it's possible to mitigate some of this in the recording stage by measuring distances, it's best practice to confirm the phase alignment during the mix stage to get the best sound.

This is not exclusive to drums, but it is easiest to demonstrate with them. Anytime you are working with multiple microphone recordings, checking the phase is a valuable tool to ensure you are getting the best mix.

There's more...

Trim isn't the only plugin with an invert phase button. In the **Channel Strip** or **EQ3** plugin window, you will notice the same symbol (ø) available. This functions exactly the same, so if you intend to use those on a track, you can opt for that instead of the **Trim** plugin.

While inverting the phase is a fundamental way to correct issues, you can also manually shift tracks forward or backward to fine-tune alignment. Zoom in very close and use the grabber tool to achieve this. You can also set the nudge value very low and use the nudge keys in a similar fashion. You can also use a **Delay** plugin to dial in an exact value to the millisecond, as long as the feedback is set to 0.

There are also plugins designed for phase alignment that work with finer controls over the phase adjustment. Audio signals and their phase alignment can be represented mathematically as a degree in a circle, with 180 degrees being fully inverted. Some plugins allow 90-degree or other increments to get exact results. As with other plugins, explore what's available and use a trial period to decide whether it's worth it in your workflow.

Using expanders to create more "punch"

Previously in this book, we explored using compressors and limiters to increase the overall loudness of a track. While you can (and should) use this for many types of instruments, using an expander to bring out the transients or attacks first will ensure the impact of the instrument is heard over the other parts captured. Once again, we'll use drums as our basis as it's easiest to demonstrate.

Two parts of a drum kit that are easiest to conceptualize are the kick and overheads. A kick drum is usually supposed to sound like a hard percussive thud. A well setup and recorded kick drum might be able to achieve this sound without further processing, but more often the microphone will capture other parts of the kit and the "ring" of the kick drum as the impact resonates throughout the internal structure of the instrument. Overheads meant to capture the cymbals also have this challenge, although you usually want to have them ring out a bit. An expander will let the sharp initial impact cut through the mix, then let the softer tones die out either quickly or slowly depending on the settings.

It might be hard to hear the difference when listening to the drum kit on its own, but using these tools definitely helps the kit cut through the mix when other instruments are added.

In this recipe, we'll take the same drum tracks from before and apply an expander to different tracks to help increase the dynamics and bring that "punch" that's needed.

Getting ready

For this recipe, you will need a Pro Tools session with a drum kit recording with separate mics/tracks for each piece of the kit. You can use the kit recorded in the previous recipe or grab a new session obtained in the GitHub project linked in the *Technical requirements* section at the beginning of this chapter.

In this instance, the Overhead tracks are easier to work with if they are in a stereo track.

Make sure, in your Pro Tools session, that the **Inserts** column is shown in the **Edit** window. You can do this by going to the menu bar and selecting **View** | **Edit Window Views** and making sure the appropriate items are checked, or by using the **Edit Window View Selector** dropdown directly above the track headers on the left side of the **Edit** window.

How to do it...

We'll add some expanders using the **Channel Strip** plugin to the kick and overheads and play with the settings to dial in the right sound. Follow along with these steps:

1. Solo the **Kick Drum** track by clicking the **S** below the track name.

2. In the **Kick Drum** track, click on the first available insert slot and select **plug-in** | **Dynamics** | **Channel Strip**.

3. In the middle of the **Channel Strip** plugin window, click on the tab named **EXP/GATE**.

4. Set the following parameters for **EXP/GATE**:

    *   **RATIO:** 1:4.0
    *   **THRESH:** -80 dB
    *   **ATTACK:** 20.0 us
    *   **DEPTH:** -22.0 dB
    *   **HOLD:** 1.0 ms
    *   **KNEE:** 0.0 dB

- **RELEASE:** `100.0 ms`

Figure 7.4: EXP/GATE settings in the Channel Strip plugin

5. Begin playback with the *spacebar* or the transport.

6. As the track plays, slowly increase the **THRESH** (threshold) knob until you hear the kick start to sound more pronounced and the resonance is reduced – you may need to play around with this setting a bit and compare how it sounds

7. Click the solo button (**S**) on the **Kick Drum** track to disable the solo and hear how it sounds with the rest of the mix.

8. Hold *command* (*control* on Windows) and click on the **Channel Strip** insert repeatedly to toggle the bypass and hear the difference, if you are happy with the sound, keep it enabled; otherwise, tweak **THRESH** to your liking (you can also adjust the **RATIO, DEPTH,** and **RELEASE** knobs to see their impact).

9. Repeat *steps 1-8* for the **Overhead** tracks, but this time set **RELEASE** to `2.5 s` (much longer).

10. Adjust the volume and panning for each track to your liking until you are happy with the mix. You can also play with adding an expander to other tracks to hear what they sound like.

How it works...

Using expanders/gates was explored in *Chapter 5* as a noise reduction tool, and the concept here is the same. Expanders work by allowing sounds that are above the threshold you set. The other parameters dictate how aggressively the increase in dynamics is applied, and how quickly or slowly it reacts to the change. For instruments like a kick drum, quick releases are usually needed to create that "tight" thud. For instruments such as cymbals, a slower release is needed otherwise the gate will sound too noticeable and be distracting.

Finding the right setting can be challenging if you don't know where to start, but I find the settings outlined in this recipe work most of the time and can be tweaked to your liking. There are also some presets in the **Channel Strip** plugin that apply EQ and other tools to the tracks as a good starting point. Try **Kick Drum Hyped** to hear what it does, but you will still need to adjust the **EXP/GATE** settings to apply what we've explored here.

There's more...

Some expanders/gates are better suited for musical scenarios. It's why I opt for Avid's Channel Strip as opposed to the Dyn3 Expander/Gate. There are other plugins you can try to achieve different results, and I especially enjoy using multichannel expanders for these applications such as WaveArts Multidynamics and Waves C4 or C6, though they are outside of the scope of this book. I recommend trying out different expander/gate plugins to hear how they react to different instruments and decide which one may work best given a scenario. As we explored here, channel strips or plugins that include a noise gate in some form can also work well.

While we only worked with drums in this recipe, try out adding a gate to different types of tracks to hear the results. Gating reverb tracks was a very common practice for drums in the 80s and will get you that iconic and nostalgic sound. Using it on vocals can also add that punch or remove reverb from a track. You can also add an expander to a submix or the entire mix on an Aux Track to liven up the sound.

Applying sidechain compression

Sidechain compression might sound like a fancy term, but the concept is fairly simple: take one track's signal, and let it control a plugin on another track. When applied to a compressor, you can think of it as "ducking" the track. An example that's easier to imagine is a podcast scenario. Let's say you have a music track that you want to lower every time someone speaks. You apply a compressor to the music track, but instead of using the music's own signal to lower the volume, it's set to the speaker's. Every time the speaker talks, the music's compressor will react and lower the volume, and when they stop talking the music will come back up in volume according to the release setting.

Musically, sidechain compression can serve two main purposes:

- To create a rhythmic "breath" or "pulse" to a track (common in electronic dance music)

- To lower certain tracks momentarily so they don't compete with others

For this recipe, we're going to use the kick drum's signal to duck a bass track so that they don't compete. This is facilitated via a signal bus, which is what is used to route Aux and Print tracks, but in this scenario, it won't be sent to an audible track.

Getting ready

For this recipe, you will need a Pro Tools session with a drum kit recording with separate mics and tracks for each piece of the kit as well as a bass track. You can use the example session downloaded from the GitHub project linked in the *Technical requirements* section at the beginning of this chapter.

Make sure, in your Pro Tools session, that the **Inserts** and **Sends** columns are shown in the **Edit** window. You can do this by going to the menu bar and selecting **View | Edit Window Views** and making sure the appropriate items are checked, or by using the **Edit Window View Selector** dropdown directly above the track headers on the left side of the **Edit** window.

How to do it...

We're going to send a kick drums signal to a bus and use that to drive a compressor on a bass track. Follow along with these steps:

1. On the **Kick** drum track, click on the first available **Send** slot and select **bus | Bus 1**.

> **Setting up busses**
>
> If you are not familiar with setting up bus tracks, check out *Chapter 4*, recipe *Routing signal paths for a mix*. In this case, we're using the default bussing setup, but you can rename **Bus 1** to **Kick Send** or something else to help you keep track of it.

2. On the send fader window that appears, hold *option* (*alt* on Windows) and click on the fader to set it to 0 dB.

3. On the **Bass** track, click on the first available **Insert** slot and select **plug-in | Dynamics | Dyn3 Compressor/Limiter**.

4. In the plugin window, at the top left, below the plugin name, you will find a symbol that looks like a key with a drop-down menu next to it labeled **"no key input"**. Click on the dropdown and select **bus | Bus 1**.

5. On the right side of the plugin window, next to the **SIDE-CHAIN** label, click the button that looks like a key.

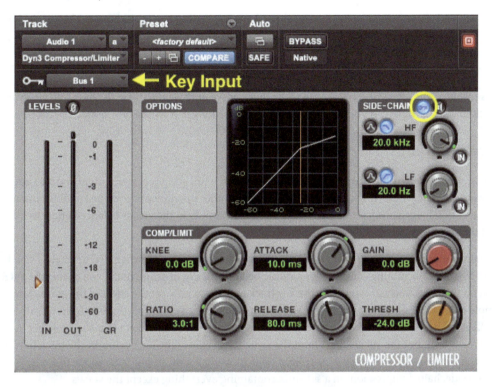

Figure 7.5: Key input selector and side-chain enable

6. Set the compressor settings as follows:

    * **KNEE:** 0.0 dB

    * **ATTACK:** 20.0 us

    * **GAIN:** 0.0 dB

    * **RATIO:** 8:1

    * **RELEASE:** 20.0 ms

    * **THRESH:** 0.0 dB

7. Start playback with the *spacebar* or the transport.

8. As the track plays, slowly bring down the **THRESH** (threshold) knob until you see the **GR** (gain reduction) meter is showing a good amount of activity (the meter is almost fully saturated when the kick plays, but doesn't stay on all the time).

How it works...

Using the Send allows you to "split" the signal and have it routed to one of the busses. This bus is then set as the key input in the compressor, and instead of the compressor using the bass signal to lower the volume, it uses the kick's. You will see other plugins also have a key dropdown and a way to select a bus, as well as a way to turn on or off the sidechain.

In this instance, the attack and release are really quick as we only want to drop the bass volume when the kick makes an impact. If you were to make the release longer, it would be a noticeable drop.

There's more...

In this example, we're only dropping one track but with a full song mix, you might want to add a compressor to other tracks that compete with the kick or other instruments. Even just a subtle amount of compression being sidechain keyed from a vocal track can help the vocals balance better over the rest of the music bed when they are singing.

Sidechain compression with a longer release and a steadier beat can also be applied to create a steady kind of rhythmic "breathing" or pulse for a music bed or series of instruments. If you enable the **PRE** button on the Send, you can even mute the key track entirely to only get the effect of the pulse. A cool example of this technique is used in the Nine Inch Nails track *Beside You In Time* from the album **With Teeth**. The track begins with a steady drone of guitars and synthesizers that suddenly begin to pulse with a beat, but the vocals don't have the same pulse effect. The kick drum in the track is very quiet in the mix, so this would have likely been done by using a send-in pre-fader mode to a bus, then applying sidechain compression on a sub-mix containing everything except the vocals.

Working with vocals

Vocal tracks require special care when mixing, as they will be the element within a song that the audience identifies and latches onto on their first listen. Of course, other instruments are important, and if your music has no vocals then the focus will be shifted elsewhere, but our ears are evolved to tune into the human vocal range and focus in on it.

There is a lot of leeway with what can be considered the "right" way to work with vocals, and if you speak to 100 engineers, you'll get 100 different answers as to what works best. The goal of this recipe is not to show you the definitive way to work with vocals, but instead to give you a taste of some of the more common vocal mixing techniques and what can be achieved with them. Once you know some of the tools and how they work, you can play around and experiment with them to discover different results.

Getting ready

For this recipe, you will need a Pro Tools session with a vocal track. Having a music bed included with it also helps as the way a vocal sounds in the mix is important. An example session has been included in the GitHub project files linked in the *Technical requirements* section at the beginning of this chapter.

Make sure in your Pro Tools session that the **Inserts** and **Sends** columns are shown in the **Edit** window. You can do this by going to the menu bar and selecting **View | Edit Window Views** and making sure the appropriate items are checked, or by using the **Edit Window View Selector** dropdown directly above the track headers on the left side of the **Edit** window.

How to do it...

We're going to take a vocal track and add some plugins to it in a specific order. We'll also send the vocal track to some Aux tracks for some added effects. Follow along with these steps:

1. In your Vocal Track's header, click the first insert slot and select **plugins | Dynamics | Channel Strip**.

2. Solo the **Vocal** track with the **S** button below the track name.

3. Begin playback with the *spacebar* or the transport.

4. In the **Channel Strip** plugin window, begin with the **EXP/GATE** tab – set **RATIO** to 1 : 4 . 0 then adjust the **THRESH** knob until it dries up the vocals but doesn't cut off the performer's voice.

5. Move on to the **COMP/LIMIT** tab and set **RATIO** to 6 : 1.

6. Adjust the **THRESH** knob until there is a good amount of compression being performed. Refer to *Chapter 5*, recipe *Using Compression to control dynamics*, if needed.

7. Disable **Solo** on the track and listen to the vocals in context with the music bed. It should sound fairly quiet in comparison.

8. While still playing back, in the **Channel Strip** plugin window, slowly increase the **GAIN** knob in the **COMP/LIMIT** tab until the vocals punch through the mix and sound pleasing to your ears.

9. Work on the EQ next. You can use the Channel Strip EQ for simplicity or add an EQ3 7-band to the next available insert slot if you prefer – the settings will be to your liking so play around with the different bands and see if increasing or decreasing frequencies helps and refer to *Chapter 5*, recipe *Using EQ to shape the tone of a sound*, for more details if needed.

10. Stop playback.

11. Click on the next available insert slot in the Vocal track header and select **plug-in | Dynamics | Dyn3 De-Esser**.

12. In the De-Esser plugin window, under the **Preset** area, click the dropdown and select **Male De-Ess HF** (select Female if using a female voice).

13. Click on the next available insert slot in the Vocal track header and select **plug-in | Harmonic | Lo-Fi**.

14. In the Lo-Fi plugin window, toward the bottom, set the **DISTORTION** knob to `0.7` and the **SATURATION** knob to `0.1`.

15. Click on the first available **Send** slot and select **new track…**.

16. Create a **Stereo Aux Track** and name it `Delay`.

17. Repeat *steps 14-15*, but name this track `Reverb`.

18. In the **Delay** track, click the first **Insert** slot and select **multichannel plug-in | Delay | Mod Delay III**.

19. In the **Mod Delay** plugin window, under the **Preset** tab, select **ModDly II – Slap**.

20. In the **Reverb** track, click on the first available **Insert** and select **multichannel plug-in | Reverb | D-Verb**.

21. In the **D-Verb** plugin window, click the **AMBIENT** and **LARGE** buttons.

22. Begin playback again.

23. In the **Vocals** track, click the first send slot (Delay) to make the send fader appear.

24. Increase the delay send level until it is to your liking.

25. Repeat *steps 23-24* but with the reverb send.

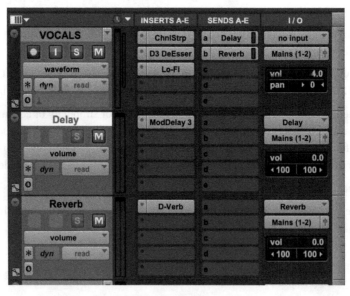

Figure 7.6: An example vocal chain

How it works...

We've applied a lot of different effects and tools to our vocal chain, so let's break it down. The chain goes in this order:

- Channel Strip:

  - Expander: Removes noise

  - Compressor: Smooths out the dynamic and increases loudness

  - EQ: Shapes the tone of the vocals

- De-Esser: removes harsh sounds typically caused by sibilance ("ess" sounds)

- Lo-Fi: Adds some distortion and "dirties" the signal a bit for character

- Delay: adds an echo

- Reverb: adds some ambient room reverberation to the track

The order these inserts are added in is important since the chain is sequential. The output of the Channel Strip goes into the De-Esser and that subsequently gets fed into the Lo-Fi Insert. If you were to add Lo-Fi first, then the expander would work less effectively due to some of the characteristics applied. This isn't to say that this is the best or only way to chain a signal path.

Some engineers prefer to have dynamic controls like compression applied at the very end of the chain, where others prefer it before EQ. Some also like both, starting with an EQ, then a compressor, then another EQ to shape it again. Placing multiple instances of a plugin to work iteratively is perfectly acceptable.

You can also use sends for further creative applications. You could send the vocal signal to another Aux track and apply more aggressive effects/filters and adjust the volume level of both tracks to create an interesting blend. I also like sending the delay signal to the reverb track in addition to the main track to give the delay some "air."

Manipulating tempos

Being able to set the tempo for a piece of music is a fundamental aspect of music production as it allows for on-the-beat editing and certain plugins to be synced perfectly. But what happens if you've received tracks or a project that doesn't have its tempo set and want to implement it?

Luckily, there are a few methods for rectifying this, and we're going to explore two of them in this recipe.

Getting ready

For this recipe, you'll need a Pro Tools session with a mixed song of your choosing imported into a stereo track. For simplicity's sake, it's best to choose one in 4/4 time (the pulse of the song goes in a 1,2,3,4 beat), though if you are comfortable with other time signatures you can choose otherwise.

You'll also want the toolbar to show **MIDI Controls**. To do this, right-click anywhere on the toolbar at the top of the Pro Tools window and check **MIDI Controls** from the dropdown.

Make sure, in your Pro Tools session, that the **Inserts** column is shown in the **Edit** window. You can do this by going to the menu bar and selecting **View | Edit Window Views** and making sure the appropriate items are checked, or by using the **Edit Window View Selector** dropdown directly above the track headers on the left side of the **Edit** window.

How to do it...

We're going to set the tempo of a song using both the tap tempo and identify beat tools within Pro Tools. We'll also add a click track to help confirm that the tempo we set is correct. Follow along with these steps:

1. Under the **MIDI Controls** tab in the toolbar, make sure the **Metronome** and **Conductor** buttons are active (the second and fourth buttons from the left).

Figure 7.7: The Midi Controls

2. Begin playback with the *spacebar* or the transport.

3. In the **MIDI Controls** tab in the toolbar, click on the number next to **Tempo** (it should be **120.0000** by default).

4. On your keyboard, repeatedly press the *t* key to the beat of the song as you listen – you should see the tempo change to match over time.

5. When the tempo has stabilized, stop pressing *t* and press *return* (*enter* on Windows) to set the tempo – this may take a few measures of the music to be accurate.

6. Stop playback.

7. In the menu bar, select **Track | New…**.

8. Create a **mono Aux Track** and name it Click.

9.   In the Click track, click on the first Insert slot and select **plugin | Instrument | Click II**.

10. Begin playback again and confirm that the tempo being created by the click track matches the tempo of the song.

11. If the tempo does not match, you can click on the **Tempo** value in the **MIDI Controls** tab in the toolbar and use the *up* and *down* arrows on your keyboard to increase or decrease it accordingly.

12. Mute the tempo track with the **M** button below the track name.

13. Press *return* on your keyboard (*Enter* on Windows) to reset the timeline insertion to the beginning of the session.

14. Begin playback with the *spacebar* or the transport and count the beats as the song plays – assuming the song is in 4/4 time, each time you count 1, increase the first number as you count (1,2,3,4; 2,2,3,4; 3,2,3,4; 4,2,3,4). When you hit 5, stop playback.

15. In the menu bar, select **Event | Identify Beat**.

16. In the **Location** field, click the first number and type in 5.

17. Click **OK**.

18. Observe the new tempo in the **MIDI Controls** window.

19. Unmute the **Click** track and start playback again to confirm the tempo is correct.

20. If the tempo changes over time, continue to listen and count the beats, stop playback after a number of bars, and use the **Identify Beat** tool again to update the value.

How it works...

There are two methods in this recipe. In the first method, we used the **tap tempo** tool to match the beat to the existing song. It makes sense that the *t* key would be used for tempo, but it's not overtly evident that you need to click on the tempo meter to enable it.

The second method is more useful in scenarios where tempos change over time. Instead of setting the tempo of the song, you are identifying where in the song the beat lies, and Pro Tools will perform the math to get the correct tempo. If I know the musicians were playing to a click track, or if the song is electronic/sequenced, I will opt for tap tempo out of convenience. If the performers were not using a metronome, then the second option will yield more accurate results.

There's more...

By changing the transport's display to **Bars|Beats**, you'll be able to easily see where in the track the beats are divided (it will show in the main ruler). You can also use the Grid mode to now edit perfectly to the beat, as all your edits and selections will always line up with those bars and beats. Combining this with shuffle mode will make sure the song is always in sync and in tempo as you edit it. As we saw with the **Click** plugin, there are also plugins that will follow the tempo of the track.

Post Production for Motion Pictures

Sound for motion pictures might be one of the most scrutinized mediums in the realm of audio. When less-than-ideal audio is present in a music composition, it can sometimes be attributed to the artist's intentions or style (such as the intentional warble, noise, and crackle in Lo-Fi). In podcasts, the journalistic tradition of radio forgives recording quality as it may not be done in an ideal space or situation. With motion pictures, however, bad sound is almost unforgivable. Audiences will accept out-of-focus images, shaky cameras, or overexposed bright spots, but present them with a soundtrack that is out of sync, distorted, or lacking in the edit, and their immersion will be broken. Motion picture audio also falls under heavy scrutiny in the broadcast world with specific quality assurance targets that will cause a project to be rejected if they are not respected.

While the challenges present with motion picture audio may seem daunting, Pro Tools provides some specific tools that greatly help overcome those hurdles. I started my post-production journey with motion picture audio in Pro Tools, and while I've dabbled with other DAWs, I still feel that it's superior in this area for my workflows (with a close second being Nuendo).

In this chapter, we'll focus on specific tools that support areas of motion picture audio such as dialogue, foley, and effects, and also how to handle changes when they come down the pipeline.

The recipes we'll cover are as follows:

- Setting up a session for motion picture projects
- Setting up a mix
- Stretching clips with Elastic Audio
- Manipulating timing with warping tools
- Preparing an **Additional Dialogue Recording (ADR)** session

- Performing an ADR session
- Conforming a session when things change

Technical requirements

The recipes in this chapter require Pro Tools Ultimate, but some workflows will still work with other versions of Pro Tools. Example files for select recipes can be found at `https://github.com/PacktPublishing/The-Pro-Tools-2023-Post-Audio-Cookbook`.

Setting up a session for motion picture projects

There are certain technical aspects of a motion picture project that are important to consider when approaching a project. This recipe will take you through the setup of a motion picture mix template and discuss the reasoning behind the choices in the setup.

Getting ready

For this recipe, you will need a blank Pro Tools session with one mono track named Beep. As this is designed for motion pictures, having video content to work with helps so an example video file with a universal leader and timecode burned into it has been placed in the sample files on the GitHub page listed under the *Technical requirements* section of this chapter.

How to do it...

We'll set up a session to follow the **Society of Motion Picture and Television Engineers** (**SMPTE**) guidelines, which are required for most motion picture projects. This includes making sure the timecode for picture start is correct, and a "2-beep" is placed. Follow these steps:

1. In the menu bar, select **Setup | Session**.
2. On the top right of the **Session Setup** window, click in the field for **Session Start** and enter `00:58:00:00`.
3. Click on the dropdown for **Timecode Rate** and select the correct **frames per second** (**fps**) according to the video you are importing (if you are using the example video, set it to `29.97 fps`).

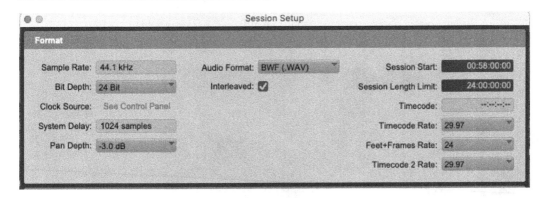

Figure 8.1: Session Setup

4.   Close the **Session Setup** window.

5.   In the toolbar, click the triangle next to the counter and select **Timecode**.

Figure 8.2: The counter

6.   Click on the counter and type in 01000000 and press *Enter/Return*.

7.   Press *Enter* on the numpad to create a marker at **01:00:00:00** – name it Picture Start.

8.   Click on the name for the **Beep** track to select/activate it.

9.   Click on the counter again and press *backspace/delete*.

10.   Type in 595800 and press *Enter/Return*.

11.   In the **Edit Selection Indicators** (directly to the right of the counter), click on the timecode next to **End**.

12. Press the left arrow on your keyboard twice to highlight the frames field of the **End** timecode.

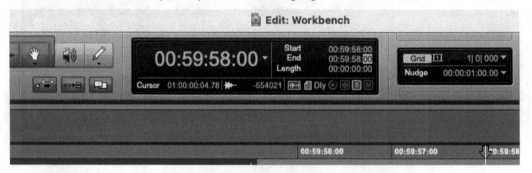

Figure 8.3: The End timecode indicator with frames selected

13. Type in 1 and press *Enter/Return* – this should highlight 1 frame of empty audio in the timeline.

14. In the menu bar, select **AudioSuite | Other | Signal Generator**.

15. The default settings should be set to a **1000 Hz Sine Wave at -20.0 dB**, this is perfect for our needs, so click **Render**.

16. In the menu bar, select **File | Import | Video…**.

17. Navigate to the sample video file provided and click **Open**.

18. In the **Video Import Options** window, set **Destination** to **New Track** and **Location** to **Spot** (the **Destination** option is only available in Pro Tools Ultimate; Studio will default to the main video track).

19. When the **Spot** dialog appears, set **Start** to 00:59:39:00.

20. Use the Select Tool and click around the video track to confirm that the timecode in the video matches the timecode in the counter. If it doesn't, set the **Nudge** value to 1 frame, highlight the video clip with the Grabber Tool, and use the nudge keys (, and .) to move the video back or forth until it's aligned.

21. Use the Grabber Tool and click on the video clip to select it.

22. Press *command + L* (*Ctrl + L* on Windows) to lock the video file, preventing it from being moved out of sync accidentally.

How it works…

The end goal of a motion picture mix is to deliver a mixed audio file (or files if you need to submit a multi-mono track) to the client. While it's possible to have your session start at 00:00:00:00, this is generally bad practice for a few reasons, the easiest to identify is ensuring the picture start is correct. Since the timecode can only be a positive value (negative time doesn't exist), the convention is to have the picture start at the 1-hour mark, 01:00:00:00. This gives the opportunity to have color bars and tone at the head of a video project (important when working with tape formats), as well as any title cards/slates with important information you may need to include about the video itself. Finally, a Universal or SMPTE leader before the picture start that includes a beep 2 seconds before the picture start will ensure sync in any scenario.

When working on a series, or multiple short videos that are part of a larger campaign or project, it's common practice to have each episode or section increment by one hour. So, episode 1 starts at 01:00:00:00, episode 2 starts at 02:00:00:00, and so forth. This can be really handy as you could potentially place multiple videos in the same Pro Tools session with all the episodes spaced apart. This provides an opportunity to reuse assets (such as musical stingers) easily if needed, as well as keep things consistent across episodes if certain signal chains are applied.

Setting up a mix

In previous chapters, we've explored how to route audio between tracks including sub-mixes and Aux Tracks. This recipe puts that all together in a direct template for setting up a mix session for a motion picture project. We'll create different tracks for Dialogue, Effects, Foley, Music, and Ambience, route them to submixes and reverb auxes, then pass that all through a brick wall limiter, and finally, a print track. While not every project might need all these tracks, it's easier to set up more than you need and subtract from it later than adding extra tracks later and making a mistake in the routing.

Getting ready

For this recipe, you will need a blank Pro Tools session. You can use the session you set up in the previous recipe or an example session linked in the *Technical requirements* section of this chapter.

Make sure in your Pro Tools session that the **Inserts** and **I/O** columns are shown in the **Edit** window. You can do this by going to the menu bar and selecting **View | Edit Window Views** and making sure the appropriate items are checked, or by using the **Edit Window View selector** dropdown directly above the track headers on the left side of the **Edit** window.

How to do it...

We'll create a bunch of tracks and route all the audio accordingly for use with a motion picture project. Follow along with these steps:

1. In the menu bar, select **Track | New...**.

2. Create the following tracks using the plus sign to add additional tracks (the names for the groups of tracks are after the dashes):

   - 1, Stereo, Audio Track – `Print Master`

   - 1, Stereo, Aux Input – `Limiter`

- 5, Stereo, Aux Input – `Submix`
- 2, Stereo, Aux Input – `Reverb`
- 3, Mono, Audio Track – `Dialogue`
- 12, Mono, Audio Track – `Effects`
- 12, Stereo, Audio Track – `Effects (Stereo)`
- 12, Mono, Audio Track – `Foley`
- 2, Stereo, Audio Track – `Music`
- 8, Stereo, Audio Track – `Ambience`

3. Press the **Create** button.

4. Hold *option* (*Alt* on Windows) and click on any track name to deselect all tracks.

5. Double-click on the track name for **Submix 1** to bring up the track info window.

6. Under the **Name the track** field, type in `Dialogue Submix`.

7. Press **Next** or *command + right* (*Ctrl + right* on Windows) to move to the next track.

8. Name the rest of the submixes: `Effects Submix`, `Foley Submix`, `Music Submix`, and `Ambience Submix`.

9. Click **OK** when you have finished naming.

10. Click on **Dialogue Submix** to select it, then hold *shift* and click on **Reverb 2** to highlight all the submix and verb aux tracks.

11. Hold *option + shift* (*Alt + Shift* on Windows) and click on the output selector in the **I/O** column on any of the highlighted tracks (second dropdown from the top below the input).

12. Select **Track | Limiter**.

13. Click on the **Limiter** track's output and select **Track | Print Master**.

14. Click on the **I** character below the **Print** track to enable Input Monitoring.

15. Click on the name for **Dialogue 1** to select it, then hold *shift* and click on **Dialogue 3** to select all of the **Dialogue** tracks.

16. Hold *option + shift* (*Alt + Shift* on Windows) and click on the output selector in the **I/O** column for one of the **Dialogue** tracks.

17. Select **Track | Dialogue Submix**.

18. Repeat *steps 15 to 17* for each group of tracks and set their outputs to their corresponding submix, keep in mind both Stereo and Mono Effects get routed to the same submix.

19. On the **Limiter** track, click the first **Insert** and select **plug-ins | multichannel plug-ins | Dynamics | Maxim**.

20. In the **Maxim** window, click the **LINK** button, then slide the threshold to **-11 dB**.

Your session is now set up and routed for mixing a motion picture project.

How it works...

This mix session follows some conventions that I've used over the years. By routing everything through a Limiter, you can ensure that the signal level doesn't overload and fail Quality Assurance according to where it's being delivered. You can also use the Limiter to adjust the overall loudness, as we've examined in earlier chapters. It is also useful to add a LUFS meter to the Limiter track (see *Chapter 6, recipe Understanding LUFS,* for more information).

Each set of tracks is set to a submix. This provides some practical advantages, such as being able to mute entire groups in one click to focus on certain areas of a mix. If you have a good balance within a group as well, you can simply make a level adjustment to a submix if it doesn't sit well in the overall mix. For example, you may start with a great Dialogue mix, but as you move on and add all the other tracks, it starts to get lost. Instead of going to each track and painstakingly rebalancing the levels, you can apply a volume adjustment to the submixes to get things sitting well again.

Finally, submixes also allow you to print stems more easily, something we've discussed in earlier chapters as well.

There's more...

I find it very handy to change the track colors as well to fit within the different groups. I might have all the effects green, or all the music light blue. I usually also define the Print track, Limiter, and Submixes into three different colors to help define things.

Stretching clips with Elastic Audio

Elastic audio is super powerful in the motion picture realm. It provides the ability to stretch and shorten the timing of a clip but does so in a non-destructive way. Instead of using AudioSuite and Time Shift, you can dynamically adjust the timing of clips according to how you want directly on the timeline. This can be used to adjust the timing of dialogue that was recorded in an alternate take, or in ADR. This can also be used to adjust the timing of effects or music. Elastic audio usually is applied in a manner that's somewhat transparent (to a degree), but it can also be used with Varispeed to slow down and speed up the pitch of tracks for creative purposes. All this can be experimented with and tested in a way that doesn't add additional artifacts through iteration, and can all be reverted easily as well thanks to its non-destructive nature.

Getting ready

For this recipe, you will need a Pro Tools session with a short clip (1 or 2 seconds) placed in an audio track. Any kind of audio clip will do, but you will notice different types of elastic audio manipulation are suited for different types of audio (noted later). Example audio files have been provided in the GitHub link provided in the *Technical requirements* section of this chapter.

How to do it...

We're going to manipulate an audio clip with elastic audio and explore the different types of elastic audio available. Follow along with these steps:

1. In the Track Header, click the grayed-out button toward the bottom that looks kind of like a thumb tack resting on a table (see *Figure 8.4*) – if you don't see it, you may need to increase the track height.

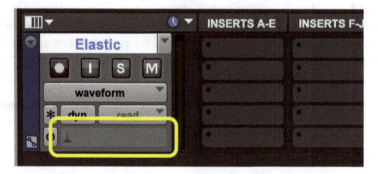

Figure 8.4: The Elastic Audio drop-down selector

2. From the **Elastic** audio selector, choose **Polyphonic**.

Figure 8.5: Elastic audio options

3. In the toolbar, click and hold on the **Trimmer Tool** to bring up the different modifiers and select the **TCE** tool (you can also cycle through the different options by pressing *F6* on your keyboard).

Figure 8.6: The TCE tool

4. Float the cursor over the right edge of the audio clip to have the **TCE** icon appear.

5. Click and drag the edge of the clip to the left to shorten it – if you have the Clip List open, you will notice that a new clip does not get generated.

Figure 8.7: Comparison of the unaltered clip (left) and the compressed clip (right)

6. Press the *spacebar* on the keyboard or play with the transport to hear the change in timing.

7. Click and drag the right edge toward the left to continue shortening it; you will notice at a certain point it turns red – this is to indicate that you've compressed it beyond the point that Pro Tools is confident it can apply the change effectively.

Figure 8.8: A clip compressed too far

8. Click and drag the right edge toward the right to stretch the clip out; again, continue stretching it to see the clip turn red to warn you.

Figure 8.9: A clip stretched too far

9. Right-click on the clip and select **Remove Warp** to restore the clip to its original length.

10. Right-click again on the clip and this time, select **Elastic Properties**.

11. In the **TCE Factor** field, type in a value of 50 and press *Return/Enter* on your keyboard.

12. With the **Elastic Properties** window still open, click and drag the right edge of the clip – notice the **TCE Factor** field within the **Elastic Properties** window updates dynamically.

13. In the **Elastic Properties** window, click on the **Pitch Shift Semi** field and type in 2, then press *Return/Enter*.

14. Preview the sound of the Pitch Shift effect with the *spacebar* or the play button on the transport.

15. With the **Elastic Properties** window still open, click on the **Elastic** audio dropdown selector in the Track Header and select **Monophonic** – observe how the **Pitch Shift** fields are now grayed out.

16. Click the **Elastic** audio dropdown selector and select **Varispeed**.

17. Either using the **Elastic Properties** window or the **TCE** tool, stretch or compress the clip and play back the clip to preview the results.

18. Click on the **Elastic** audio dropdown selector in the Track Header and select **X-Form**.

19. Use the **TCE** tool and stretch or compress the audio clip; observe how the clip is initially grayed out, then becomes rendered and playable after some time.

How it works...

Elastic audio functions in a similar way to the TCE tool on its own. You can stretch or compress audio clips to change their speed, but with Elastic Audio disabled, each clip is printed/rendered, meaning that the process is destructive, much like AudioSuite. Multiple time changes using destructive tools will result in worsening audio artifacts, much like generational loss with analog media. With Elastic Audio enabled, the clips can be pulled and squashed, and changed dynamically without changing the original clip. You can adjust the timing many times without affecting the overall quality, as it will always reference the original track. And at any point, you can always revert to the original by removing the Warp.

Each elastic audio type has its advantages with different use cases.

Polyphonic

Polyphonic refers to any audio source with more than one "voice" or instrument. This could be a music bed, or multiple instruments playing. This sometimes also works well for dialog as human voices tend to work in different simultaneous tonalities.

Rhythmic

This is for percussion and drum tracks that have a pulse or rhythm. This type of elastic audio will try to respect the transients and warp the timing between hits as opposed to the hits themselves.

Monophonic

This is for single-voiced instruments that are generally single-toned. Think of instruments such as a saxophone or flute, where multiple notes cannot be played at the same time. Human voices that are singing also work well in this mode. Instruments such as guitars or pianos, where multiple notes overlap or are playing chords, tend to not work well. You cannot Pitch Shift in this mode.

Varispeed

This is the "original" speed change method, where increasing or decreasing the playback speed also affects the pitch. Think of how tape recorders worked, and Varispeed was a common tool used to change the pitch of a recording. I personally love using Varispeed for creative effects, such as slowing down or speeding up animal noises for creature effects.

X-Form

X-Form is an advanced algorithm for retiming clips that is supposed to provide the best results to mask any work done to the timing of a clip. As such, it requires that clips be rendered every time a change is made. When testing on a short clip, it might be very quick, but longer clips can take quite a while depending on the speed of your system. This can be very frustrating when trying to make minuscule adjustments to a clip in order to match the timing to something happening on screen, for example.

The best workflow I've found is to use Polyphonic for real-time adjustments of the clips on my timeline, then switch it over to X-Form when the work is finished.

Manipulating timing with warping tools

In the previous recipe, we learned how to use the TCE tool along with Elastic audio to adjust the timing of whole clips. Now, we're going to adjust the timing *within* a clip. The warp tools allow you to stretch and squash sections of the waveforms inside of an audio clip. This can be used for precise alignment of timing clips to match dialogue (either alternate takes or ADR), or this can be used as a creative effect to warp and distort the audio. Since it uses Elastic Audio, it is also non-destructive and can be reverted at any time.

Getting ready

For this recipe, you will need a Pro Tools session with a short clip (1 or 2 seconds) placed in an audio track. Example audio files have been provided in the GitHub link provided in the *Technical requirements* section of this chapter.

How to do it...

We're going to explore how Warping audio works with both the Analysis and Warp views. Follow along with these steps:

1. In the Track Header, click on the **Elastic** Audio selector and choose **Polyphonic**.

2. Click the View Selector dropdown (below the solo/mute buttons) and select **Analysis** – this view, along with Warp, is grayed out when Elastic Audio is not enabled.

3. Observe how the clip appears in the new view; lines have been added at transient points, or where Pro Tools believes a syllable is separated.

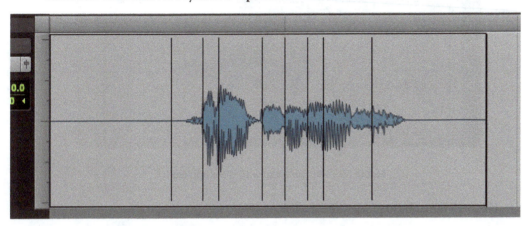

Figure 8.10: A clip in the Analysis view

4. Use the **Grabber Tool** to click and drag the lines left and right to better line up with what you believe are the correct syllables in the clip – if using the Smart Tool, float your mouse cursor on the bottom half of the clip to change it to the slider tool.

5. Hold *option* (*Alt* on Windows) and click on a line to remove it.

6. Hold *command* (*Ctrl* on Windows) and click anywhere within the clip to add a line.

7. In the Track Header, click the view selector and change it to **warp**.

8. Hold *control* (*Start* on Windows) and click anywhere on the clip to add an anchor point – you can use the lines added prior as a guide.

9. Add three anchors to the clip.

10. Make sure the clip is not selected (click outside of it if it is) and click and drag the middle line to adjust it.

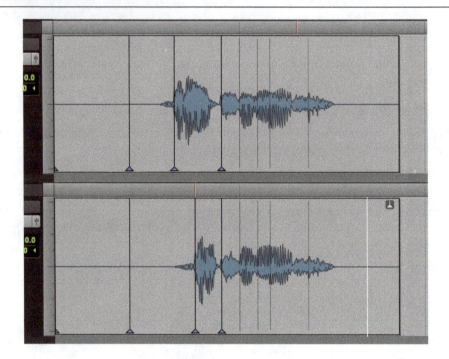

Figure 8.11: An audio clip with Warp applied

11. Play back the audio to preview the change.

12. Move the rightmost marker to see how the end of the clip moves.

13. Experiment with different anchors and Elastic Audio types to hear what the differences are.

How it works...

Using Warp tools is an extension of Elastic Audio. You can achieve a somewhat similar result by adding edits to a clip and stretching out the resulting clips, but it is cumbersome, and achieving accuracy with that type of editing is challenging. Using the Analysis view provides you with the opportunity to set or remove points you wish to use for warping. You can also add points directly within the warp tool if you know where you want to stretch and squash things. The different types of Elastic Audio and their ideal use cases still apply, so read the previous recipe if you want to see how they are implemented.

The warp tools can be used for corrective purposes, and to adjust timing for effects and dialogue, but they can also be used in creative ways, especially when Varispeed is applied.

Preparing an Additional Dialogue Replacement (ADR) session

ADR is one of the open secrets of motion pictures. The ADR process has been showcased and publicized many times, even being one of the defining moments in the movie *Singing in the Rain*. In practice though, when done well, it's transparent. The audience isn't aware that the dialogue they are hearing wasn't captured on set, and in many cases, not performed by the person on screen. When it comes to preparing an ADR session, there are a few steps you can take in Pro Tools to keep the workflow running smoothly, which is always beneficial when working with talent. You can also apply these workflows to dubbing projects, where the focus is on replacing entire lines with a different language.

Getting ready

For this recipe, you will need a blank Pro Tools session. While the process of ADR is designed to work specifically for projects with video, we will be exploring the tools and techniques that can be applied with or without one in your project. Make sure the counter is set to **Timecode**.

How to do it...

We're going to mark a few places on our timeline where we want to try and record an ADR line, and create beeps for the talent to help determine when they should start their line. Follow along with these steps:

1. Add a few markers to your project – make sure the first marker is at least 3 seconds into the project.

2. Create a new mono audio track and name it Beeps.

3. Press *Return/Enter* to bring the playhead to the start of the project.

4. Click on the name of the **Beeps** track to select it.

5. In the **Counter** window, click on the **End** counter, and then click on the frames section (the two numbers furthest to the right).

Figure 8.12: The End counter incremented one frame

6. Press *up* on your keyboard to increment the **End** counter by one frame.

7. Press *Return/Enter* and notice that the **Length** counter is now one frame long; a selection matching this should be highlighted in the track.

8. In the menu bar, select **AudioSuite | Other | Signal Generator**.

9. Press **Render** to create a one-frame beep.

10. Change the edit mode to **Spot** and activate the **Grabber Tool**.

11. Hold *option* (*Alt* on Windows) and click on the **Beep** audio clip you created (you may need to zoom into it first).

12. In the **Spot Dialog** window, click on the **Start** field and click on the seconds section of the counter (third from the left).

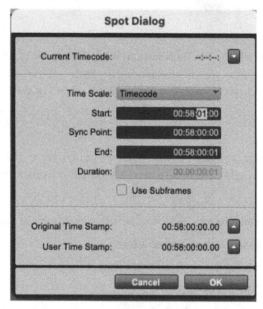

Figure 8.13: Spot Dialog with the seconds counter selected on the Start point

13. Press *up* on the keyboard to increment the seconds counter.

14. Press **OK**.

15. Repeat *steps 11* to *14* on the new **Beep** audio clip that was created.

16. Press *Return/Enter* on your keyboard to return to the start of the project.

17. In the **Counter** window, click on the **End** counter, and then click on the seconds section.

18. Increment the seconds value with *up* on your keyboard to 03 or type it in.

19. In the menu bar, select **Edit | Consolidate Clip**.

20. Add a Mono Audio track for recording takes.

How it works...

Preparing a session for ADR will help move things along during the recording phase. Markers are useful since you can bring up the memory locations to easily navigate through them, and you can use shortcuts as well (more on this in the next recipe). Labeling them with the actual line being performed is not always the best route as it can make it hard to read during the session, and most talent prefers to have the lines they need to perform in an easily readable format (on a separate screen or printed, for instance).

The Beep track is very useful for getting the talent to get the rhythm and the timing of the performance correct. By using beeps that follow a consistent beat, it functions as a countdown of sorts, and they can "feel" where the line should start by the implied fourth beat. Having a single clip that can be easily moved around also helps line things up quickly (again, we'll see this in action on the next recipe).

When it comes to naming the record track, it's up to you if you want to simply name it `Record` or `ADR`. I like to name it the character's name appended with `_ADR` to make it easier to identify in the session later, as the recorded clips will have that name applied to them.

There's more...

Markers aren't the only way to lay out the cues for talent. You can create a track and generate empty clips using the **Consolidate Clip** tool. Just like with Markers, you can give it a more generic name such as `Cue 1` or `Line 1` but due to the way clip names are drawn on the screen, I tend to find writing the line out here is more functional compared to the way markers work. If you use this method, you can navigate between the clips using the *L* and '*keys on your keyboard, or in the **Clip List**. If you opt for the Clip List, prepending the names with a numerator will make them easier to find (`001_Cue1`, for example).

Beeps are used to help provide an audio cue for talent, but you can also explore using other tools to help provide a visual cue too. EdiPrompt from *Sounds In Sync* is a popular tool that provides this along with other visual overlays to help talent in their ADR performance. You can check out this product at `https://www.soundsinsync.com/products/ediprompt`, though be advised it is geared toward studios and professionals so the prices are reflective of that. Sounds in Sync also provides other tools such as EdiCue to help organize ADR sessions, which includes the ability to export Cue sheets from Pro Tools for easier management of ADR sessions.

One of Sounds in Sync's more affordable tools is EdiMarker, which will take a text or Excel file and convert it into a Pro Tools session that can be used to import markers into your ADR session. This is useful if you work with a team and want to share Cue information across departments in a text-based form. For example, the dialogue editor working in conjunction with the director and sound supervisor might identify moments in the project where ADR is needed. Instead of trying to pass along Pro Tools sessions, the timecodes can be written in a text file for easier collaboration.

Performing an ADR session

Now that the session has been prepared, let's look at some of the ways you can quickly navigate around an ADR session while working with talent. Knowing these techniques can make things move more efficiently and save frustrations within a session. Many thanks to ADR engineer Christopher Johnson for demonstrating some of these techniques.

Getting ready

For this recipe, you will need a Pro Tools session prepared for ADR with Markers, a Beep track, and a Record track. You can reference the previous recipe to create this. This workflow uses hotkeys, so ensure that Keyboard Focus is set to the **Edit** window (the small box with **az** in the top right).

How to do it...

We'll navigate quickly between markers, place the Beep tracks accordingly, and record multiple takes while keeping them organized. Follow along with these steps:

1. In the menu bar, select **Window | Memory Locations**.
2. Click on a Marker to move the timeline insertion to that location.
3. Select the Grabber Tool, hold *control + command* (*Start + Ctrl* on Windows), and click on the **Cue Beeps** clip in the timeline – this will snap its **End** point to the marker location.

    Here's a different way to move the Cue Beeps into place for those who like to use keyboard commands more:

    I.   Use the **Grabber Tool** to select the Cue Beeps clip, or click on it in the Clip List.

    II.   Copy the clip using *c* on your keyboard.

    III. To quickly move the play head to a Marker location, use this keyboard sequence on the numpad: . *marker number* . For example, to move to marker number 2, on the numpad, press ., then *2*, then . again.

    IV.   Press *v* on your keyboard to paste the Cue Beeps clip.

    V.   Press *k* on your keyboard to move the clip's **End** point to the marker.

4. Press *L* to move the cursor to the start of the Beeps clip.
5. Disable **Insertion Follows Playback**; you can toggle this with *n* on your keyboard.
6. Record **Arm** the ADR track by clicking the red circle button below the track name.
7. In the menu bar, go to **Pro Tools | Preferences...** (**Edit | Preferences...** on Windows) and click on the **Editing** tab.

8. Make sure **Send Fully Overlapped Clips to Available Playlists While Recording** is checked to enable it.

9. Close the **Preferences** window.

10. Begin recording by either using the transport or pressing *3* on the numpad.

11. Stop recording with the *spacebar* once a take has been recorded (for demo purposes, just give it a few seconds after the beeps have played to simulate a line being recorded) – the play head should move back to the start of the Cue Beeps clip; if not ensure **Insertion Follows Playback** is disabled.

12. Begin recording again and stop it once more; you will see that when the clip stops, a *down* arrow appears over the clip and the arrow next to the track name turns blue.

13. Click the view selector dropdown on the record track and select **playlists**.

How it works...

In my experience, being able to move quickly between cues during an ADR session tends to produce a more successful result in the end. There's a disconnect between what happens in the booth and behind the board, and if you're stumbling as you move between cues, it's not always apparent what is causing the delay.

Markers are great as you can use both the memory locations and their corresponding numbers with the numpad to navigate. Using the keyboard shortcuts to quickly bring the beeps to the cue also helps immensely, and by using one of the methods listed previously, you can ensure they always line up at the start of the line every time.

Playlists are a powerful option for recording multiple takes. In some scenarios, you may prefer to simply add many tracks and move the clips down, but that adds time to the process, and having an actor that's able to redo a line essentially immediately can be very powerful in an ADR session.

Playlists work a bit like track groups, with the exception that only one track can be active at any time. You can still take pieces of different takes and paste them into the main playlist, so there are more advantages to using playlists in my opinion.

Conforming a session when things change

Probably the most contentious part of any post-audio workflow is the dreaded re-conform. There was a point in time when a picture had to be locked before it was delivered to different departments, as the technology would not allow for changes during the post process, but as technology advanced, more and more productions began practicing "soft-locks," where changes would be implemented after delivery.

This can be especially frustrating in the world of audio as we treat the sound separately from the picture and have clips overlap and extend beyond the bounds of the image's timing. Changing or removing just a few frames can have a massive impact on the audio department.

While this is not ideal, it is the reality of current production practices, and in certain scenarios, can be beneficial to all aspects of the project. We'll look at how you can manually compare OMF/AAF files within Pro Tools and match work performed to changes, and discuss some of the tools out there that can help if this is something you have to deal with often.

Getting ready

For this recipe, you will need a Pro Tools session with content imported from an OMF or AAF file, along with an additional OMF/AAF file where changes were made to the project. A sample Pro Tools session and accompanying OMF file have been provided in the GitHub project for this book, linked in the *Technical requirements* section of this chapter.

This workflow uses hotkeys, so ensure that Keyboard Focus is set to the **Edit** window (the small box with **az** in the top right).

Make sure the **Track List** is visible/expanded.

How to do it...

We'll import session data from an OMF that conflicts with our current Pro Tools session and then adjust our project to conform to those changes. Follow along with these steps:

1. In the menu bar, select **Track | New Track…** and create a mono Aux Track named - - -.
2. Right-click on the newly created Aux Track and select **Deactivate** – this track will act as a separator to help determine what changes have been made.

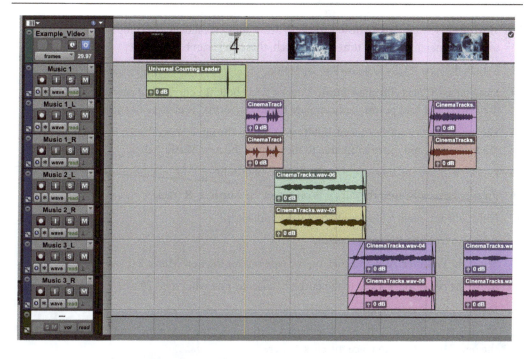

Figure 8.14: A separator track placed at the bottom

3. In the menu bar, go to **File | Import | Session Data…**.

4. Navigate to the `Version_2.omf` file and select it to be imported.

5. Leave the default options already selected and click **OK**.

OMFs and broken links

The OMF provided was exported with Adobe Premiere 2023, and might cause a warning that video files are missing. If you notice this, your best option is to select the **Manually Find & Relink** option and navigate to the OMF folder. You can find more information on this topic in *Chapter 2*, recipe *Relinking broken audio*. For these files, you may need to disable the **Match Duration** option and ignore the warning; this is a quirk of working with Premiere OMFs that have been trimmed sometimes.

---

6. Use the Grabber tool and select the Video file in the timeline

7. In the menu bar, go to **File | Import | Video…**.

8. Navigate to the **Version_2.mp4** video file and select it to be imported.

9. Set **Main Video Track** as **Destination** and change **Location** to **Selection**.

10. Use the **Grabber Tool** and select the first audio clips in the **Music 1** tracks and make note of their **Start** and **End** timecode in the counter.

11. Move the newly imported **Music 1** tracks to below the current ones and compare the waveforms; since they match and their timecodes match, these clips can be left alone.

12. Return the new tracks to below the separator.

13. Move on to the next clip in the **Music 2** tracks, select both the current and new edits. and identify their **Start** and **End** points, and compare their waveforms, as in *steps 14* and *15*.

14. The new clip has been shortened, but its start time is the same, so we'll need to trim it – activate the **Select** tool and move to an area to the right of the new **Music 2** clips.

15. Press *L* to move the Insertion Line to the end of the clip.

16. Press *p* repeatedly until the cursor is on top of the **Music 2_R** track.

17. Hold *shift* and press *p* to move to highlight both clips.

18. Press *s* to trim the clip to the cursor.

19. Select the **Grabber Tool** and click on the fade out for the new **Music 2** track.

20. Move the cursor once more to the current tracks by following *steps 20* to *21*.

21. Press *f* to fade out the clip.

22. Moving on to **Music 3**, the new clip begins earlier – use the **Select** tool and click to the left of the new **Music 3** clip to place the cursor there.

23. Change the edit mode to **Shuffle**.

24. Press ' to move the cursor to the beginning of the new **Music 3** clip.

25. Press *p* repeatedly to move it up to the current **Music 3_R** clip.

26. Hold *shift* and press *p* to extend the cursor to both **Music 3** tracks.

27. Hold *shift* and press ' to select the empty area before the clip.

28. Press *delete/backspace* to shift the clips left and line them up.

29. Continue using the previous different methods to compare the clips and align them for the rest of the project.

How it works...

This process might seem tedious, and it is. Moving, nudging, and trimming clips to match updates is time-consuming and can be mind-numbing, but if you are able to use some of the tricks just shown, you can get things lined up somewhat quickly. You might assume that you can simply import new clips into existing tracks, but if you made any edits or automation changes, that could get overwritten and lost. It's also useful to create a group for the current edit too; that way, you can move things in unison if needed. While not always possible, it's worthwhile to negotiate revision costs and timelines with potential clients to account for the extra effort required when needing to re-conform an edit after changes are made.

There's more…

We played with a fairly simple project, but when projects become large and complex, you may need the assistance of a third-party tool to help assist with the conforming process. Next are a few options you can look at.

Matchbox by CargoCult

This has the ability to compare OMFs, AAFs, EDLs, XMLs, and even the video file to find changes made. Once the changes are identified, it can make the changes to your session project by copying/cutting and pasting the clips to a different area within the tracks. It can also provide a list of changes if you prefer to look over things manually.

EdiLoad by Sound in Sync

This is designed to help sound editors with loading location sound to the scenes within a project, and it also has a compare function to help identify changes over tracks. This can be used with its Pro-Tools re-conform tool to shift tracks to their new locations.

9

Spoken Word and Podcasts

Whether you are producing audiobooks, podcasts, or read-out-loud stories, spoken word projects often have to produce a lot of content with a limited amount of time and resources. While some projects, such as audio dramas, might be more reflective of what is experienced in a typical motion picture workflow, it's more common to see longer-form work that benefits from having less manual work performed. For example, in a film that might run 90 minutes, it's expected that the re-recordist mixer uses a light amount of dynamic control and manually adjusts the volume to produce a more "natural" sounding result. When listening to a round table format podcast, the listener expects a more "radio broadcast" type sound, where the voices are heavily compressed and consistent across the hosts. It's also common for this kind of content to have quicker turnarounds and smaller budgets, so finding efficiencies will make this work more sustainable and, in some cases, more bearable to complete. We'll look at some of the techniques I've adopted for this form of media to make things go faster while not sacrificing quality.

We'll examine the following recipes in this chapter:

- Setting up a session for spoken word
- Mixing a spoken word project
- Using edit points to remove errors
- Editing during playback
- Editing while recording
- The Punch and Roll method to record voice
- Using (reverse) noise gates to fill dead air

Technical requirements

For this chapter, you will need at least Pro Tools Studio.

Setting up a session for spoken word

While spoken word projects can vary drastically, there are some common goals for how to approach the session and how to set it up to maximize efficiencies. For example, with a motion picture project, you might limit the number of dialogue tracks you create with the understanding that there are so many possible audible differences from scene to scene (and even shot to shot) that it would be impractical to have a dedicated track per person speaking. Instead, you rely on automation or clip effects to adjust the EQ and other parameters of the audio clips and have a relatively low track count. However, in spoken word and podcasts, it's expected that most of the dialogue and voices recorded will be within the same space and under similar audio conditions. Even in investigative journalism, the producer will typically sit down with the interviewee to discuss the topic with them. It's more practical to create dedicated tracks for each person and constrain each voice to that track (thanks to Rob Byers for sharing that philosophy with me). We'll set up a session with this in mind for this recipe.

Getting ready

For this recipe, you will need a blank Pro Tools session to start. Make sure that the **I/O** column is shown in the **Edit** window. You can do this by going to the menu bar and selecting **View | Edit Window Views** and making sure the appropriate items are checked, or by using the **Edit Window View selector** dropdown directly above the track headers on the left-hand side of the **Edit** window.

How to do it...

We're going to create a multi-track session that will have many voice tracks, along with some common tracks that are seen in different spoken word projects. We'll also create a few submix tracks to help with the mix. Follow along with these steps:

1. From the menu bar, select **Track | New...**.

2. Add the following tracks with the names provided:

   - 1 Stereo Aux Input: `Master Mix`

   - 1 Stereo Aux Input: `Dialogue Submix`

   - 1 Stereo Aux Input: `Background Submix`

   - 12 Mono Audio Track: `Voice`

   - 2 Mono Audio Track: `Ambience Mono`

   - 2 Stereo Audio Track: `Ambience Stereo`

   - 2 Mono Audio Track: `Clips Mono`

   - 2 Stereo Audio Track: `Clips Stereo`

   - 2 Stereo Audio Track: `Music:`

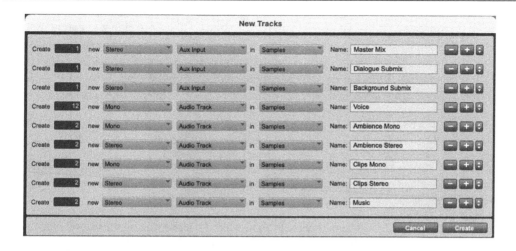

Figure 9.1 – Track setup

3. Hold *Option/Alt* and click on any track to deselect them all.

4. Hold *Command* (*Ctrl* on Windows) and click on the track names for both the **Dialogue Submix** and **Background Submix** tracks.

5. Hold *Shift + Option/Alt* and click on one of highlighted track's **output** dropdowns (the second dropdown in the **I/O** column).

6. Select **Track | Master Mix**.

7. Click on the track name for **Voice 1** track to select it.

8. Hold *Shift* and click on the track name for the **Ambience Stereo 2** track to highlight all the voice and ambiance tracks.

9. Hold *Shift + Option/Alt* and click on one of the selected track's **output** dropdowns.

10. Select **Track | Dialogue**.

11. Click on the track name for **Clips Mono 1** to highlight it.

12. Hold *Shift* and click on the track name for **Music 2** to select all the other tracks.

13. Hold *Shift + Option/Alt* and click on one of the selected track's **output** dropdowns.

14. Select **Track | Background Submix**.

How it works...

The setup for this session is geared more toward podcasts but can be applied to other spoken word projects. As mentioned at the start of this recipe, we want to be able to have one voice track for every person who speaks in this project. If you have a relatively simple project with only two people speaking, then having 12 voices won't be necessary. If one of the speakers changes spaces (perhaps there is an in-studio portion and an on-location one), then you should dedicate one track per space or setup.

Routing all the voice and ambiance tracks to a master submix will become more apparent in the next recipe. We want an easy way to control dynamics on all the voice tracks together. Similarly, by creating a separate submix for the other elements that may compete with the voices, we can use volume automation to lower those parts when someone's voice is present. We can also apply sidechain compression to "duck" the audio automatically, which will be demonstrated in the next recipe.

You do not want to have the ambiance drop when a person is speaking; it should be consistent and part of the space the person is speaking in so that it gets routed to the same submix as the voices.

Depending on the project, ambiance and other tracks (such as clips from other broadcasts or shows) might be in mono or stereo, hence why we created both. Music is almost always in stereo, though, so I've left those tracks as-is.

Feel free to add or remove tracks from this template as needed. Depending on the project, you may need tracks for effects, for example.

Mixing a spoken word project

Now that we have a session set up with the tracks needed for most spoken word projects, let's move on to mixing the project. Retaining the mindset that we should create a more "hands-off" approach to the mix, we're going to apply more aggressive dynamics control to the voice tracks and a master "catch-all" to the Dialogue Submix. We'll also apply some sidechain compression to the Background Submix so that it will automatically duck the music or other parts when someone talks.

Getting ready

For this recipe, you will need a blank Pro Tools session with multiple tracks routed appropriately for a spoken word project (check the previous recipe for instructions on how to do so).

This recipe uses the YouLean Loudness meter to measure LUFS. You will need it installed to follow through the last steps of this recipe. Check out *Chapter 6*, recipe *Understanding LUFS*, for details.

Make sure that the **Inserts** and **Sends** columns are visible in the **Edit** window. You can do this by going to the menu bar and selecting **View | Edit Window Views** and making sure the appropriate items are checked, or by using the **Edit Window View selector** dropdown directly above the track headers on the left-hand side of the **Edit** window.

How to do it...

We're going to add specific inserts to different tracks so that they're suited for spoken word projects. Follow along with these steps:

1. In the **Voice 1** track's Inserts column, click on the first insert slot and select **plug-in | Dynamics | Channel Strip**.

2. Click on the **COMP/LIMIT** tab in the middle of the Channel Strip plugin window.

3. Set the different parameters to the following settings:

    - **RATIO:** 10.0:1

    - **THRESH:** -22.0 dB

    - **KNEE:** 0.0 dB

    - **ATTACK:** 100.0 us

    - **RELEASE:** 100 ms

    - **DEPTH:** -36.0 dB

    - **GAIN:** 8 dB

4. Under the **EQ/FILTERS** area at the bottom of the plugin window, click on the second to last tab at the bottom that says **FILT 1**.

5. Click the **On/Off** button (this looks like a power button, with a circle with a line going vertically through it at the top) to activate it.

6. At the bottom left, make sure **High Pass** is selected (the first button on the left).

7. Set **FREQ** to 120 Hz and **SLOPE** to 24 dB/O:

Figure 9.2 – Channel strip settings for spoken word

8. In the **Voice 1** track's Inserts column, click on the second insert slot and select **plug-in | Dynamics | Dyn3 De-Esser**.

9. At the top of the De-Esser plugin window, click on the dropdown below the **Preset** column that says <**factory default**> and select **Female De-Ess HF**.

10. Hold *Option/Alt* and then click and drag the Channel Strip insert to the **Voice 2** track's first insert slot to duplicate it – do the same for the De-Esser, inserting it into the second insert slot of **Voice 2**.

11. Repeat *step 10* to duplicate the plugins for all the available Voice tracks.

12. In the **Music 1** Inserts column, click on the first available insert and select **multichannel plug-in | Dynamics | Dyn3 Compressor/Limiter**.

13. Set the different parameters to the following settings:

   - **RATIO**: 8.0:1
   - **THRESH**: -20.0 dB
   - **ATTACK**: 5.0 ms
   - **RELEASE**: 160 ms
   - **GAIN**: 0 dB

14. Click on the second insert slot in the **Music 1** track and select **multichannel plug-in | EQ | EQ3 7-Band**.

15. Set the different parameters to the following settings:

   - **LMF**:
     - **FREQ**: 400 Hz
     - **GAIN**: -8.0 dB
   - **MF**:
     - **FREQ**: 1000 Hz
     - **GAIN**: -3.0 dB
   - **HMF**:
     - **FREQ**: 2.00 kHz
     - **GAIN**: -8.0 dB

16. Hold *Option/Alt* and then click and drag the inserts in **Music 1** to **Music 2** to duplicate them.

17. In the Dialogue Submix track's Inserts column, click on the first insert slot and select **plug-in | Dynamics | Dyn3 Compressor/Limiter**.

18. At the top of the De-Esser's plugin window, click on the dropdown below the Preset column that says **<factory default>** and select **Brickwall**.

19. In the Sends column for the Dialogue Submix track, click on the send slot and select **bus | Bus 1**.

20. Hold *Option/Alt* and click on the fader for the Send bus that appeared to set it to 0 dB.

21. On the Background Submix, click on the first insert track and select **plug-in | Dynamics | Dyn3 Compressor/Limiter**.

22. In the plugin window, click the dropdown next to the key that says **no key input** and select **bus | Bus 1**.

23. In the **SIDE-CHAIN** area toward the top right, click the button that looks like a key.

24. Set the different parameters to the following settings:

    - **RATIO**: 20.0:1
    - **THRESH**: -24.0 dB
    - **ATTACK**: 5.0 ms
    - **RELEASE**: 500 ms
    - **GAIN**: 0 dB

25. In the **Master Mix** track, click the first insert slot and select **multichannel plug-in | Dynamics | Maxim**.

26. Click on the second insert slot in the **Master Mix** track and select **multichannel plug-in | Sound Field | YouLean Loudness Meter**.

How it works...

A lot is going on in this recipe, so let's break it down into sections and go over the different plugins and what they are doing.

Voice tracks

We have added some basic plugins to the voice tracks, with the Channel Strip doing most of the work. The big thing to note is the more aggressive settings used in the compressor area. Whereas with music and motion pictures, you may opt for more gentle ratios such as 2:1 or 3:1, we have it cranked up to 10:1. This more aggressive approach means you can rely on the compressor to do more of the work and focus less effort on volume automation.

Music tracks

We applied a compressor to the music tracks to keep them below a certain level, although most recorded and released music is already heavily compressed, so you may not want to add one. The EQ we added to the track is an easy way to prevent frequencies within the music track from clashing with the voice tracks. Most human voices range from 200 to 2,000 Hz, so I target that range and lower it. If there are music clips that are playing with no voice tracks over the top of them, I like to put them in a separate track to keep it easier to organize and not rely on automation.

Dialogue Submix

The Dialogue Submix serves two purposes. First, it applies a limiter to keep control of the voices that might get passed to their compressors. Second, it sends a full signal to a bus to be used for sidechain compression on the Background Submix.

Background Submix

While the EQ on the music helps, it can still be useful to use a side-chained compressor to "duck" the other signals. Using the bus being sent from the Dialogue Submix, we applied a compressor with a slow release. This provides a more natural sound, as if you are riding the fader up and down manually as someone speaks. If you do not like the way this sounds, it's still useful to have a submix to apply volume automation for raising and lowering other sounds as needed.

Master mix

Similar to other methods we explored in the previous chapters, using a limiter such as MAXIM combined with the YouLean Loudness meter can provide a simple yet effective way of hitting your loudness targets for delivery. Notice that there is no print track on this session. For the most part, I rely on offline bounces when completing spoken word projects as they are faster than real time. In previous chapters, I discouraged using this method, but for spoken word, the trade-offs (potentially less accurate mixes) are worth the time savings.

There's more...

The signal chain I laid out here did not go over Noise Reduction. You can use the Channel Strip's Gate/Expander to reduce noise, or dedicated plugins such as iZotope RX, Waves Clarity, or WaveArts Mutlidynamics to clean the sound before dynamics are processed. I also didn't demonstrate any EQ tools within the Channel Strip besides the filter – it will be up to you and your ears to determine what changes need to be applied for any given situation.

The Channel Strip also has a **Trim** level slider that can be used as an overall loudness adjustment before any effects in the signal chain are applied. If your original signal is too loud and the compressor is simply squashing everything, you can bring the signal down by lowering the **Trim** level. Conversely, if the signal is too low and the compressor is not activating, you can increase the **Trim** level. I often use **Clip Gain** to adjust the loudness clip to clip in place of this.

Multi-band Compressors can also be very effective as the last step in the Dialogue Submix. Instead of applying a hard limiter across the entire signal, you can target specific frequency bands, such as the lower more "boomy" tones, and the higher and harsher tones usually targeted by a De-Esser. It's just one more method of assurance to help in the effort.

Finally, tools such as Waves Vocal Rider are very handy for spoken word projects. Instead of relying on compression, or manually adjusting the levels with volume automation, Vocal Rider will adjust the volume to a set target level, similar to how a person riding the fader would. While it's not needed for every project, it can be handy in many situations.

Using edit points to remove errors

Depending on the recording process, you might find yourself with a voice track that has had errors and mistakes removed, mistakes marked on the audio track with a clicker, or no markings whatsoever. When working with voice tracks that have no guidance for editing, being able to add edits quickly as you listen to playback can be very helpful. This recipe will demonstrate one way you can edit as you listen to a playback, then go back and remove the errors after you perform a pass. There are other methods to do similar work, but this practice is still helpful for certain scenarios.

Getting ready

For this recipe, you will need a Pro Tools session with an audio track placed in a track. If you want to try this out on a practical example, then having a recorded voice with errors can help. One has been provided in the GitHub project for this book. Otherwise, an empty audio clip or any audio clip will do.

The commands in this recipe require **Keyboard Focus** to be active in the **Edit** window. This can be done by clicking the small **az** icon at the top right of the **Edit** window or pressing *Command + Option + 1* (*Ctrl + Alt + Numpad 1* on Windows).

You should also ensure that **Tab to Transients** is disabled. You can do this in the **Options** menu or by clicking the leftmost button under the tool selector:

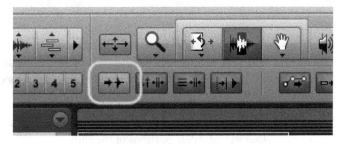

Figure 9.3 – The Tab to Transients button

How to do it...

We're going to play back some audio and add edits to it as we listen. Once the edit is complete, we'll remove the sections containing errors. Follow along with these steps:

1. Make sure you are in **Shuffle** edit mode. You can do this by pressing the **Shuffle** button at the top left of the **Edit** window or by pressing *F1* on your keyboard.

2. From the menu bar, select **Options | Edit Window Scrolling | Page**.

3. Activate the **Select Tool** property and click on the track you wish to edit at the place you wish to start the playback.

4. Press the spacebar or start the playback with the transport.

5. Listen to the track. When you identify an error or mistake, press the *down* arrow on your keyboard to move the edit insertion point to the playhead – let the playback continue, don't stop it.

6. Press *B* on your keyboard to split the clip at the insertion point. The playhead will continue moving past this point.

7. Continue listening and watching the waveforms. When you see the playhead approach another section of voice, press *down* and then *B* on your keyboard to add another edit.

8. When you're done listening to the clip, stop the playback by pressing the spacebar.

9. Press *L* on your keyboard to move the insertion point to the previous edit.

10. Hold *Shift* and press *L* to select the clip that will be removed.

11. Press *Backspace/Delete* on your keyboard to remove the clip. This will shift the next clip back.

12. Now, you need to fine-tune the edit point. You can do this with the **Trimmer Tool** property and dragging the tail of the clip to the left of the edit point, before the mistake was made.

13. Add a short crossfade to the edit point to smooth the transition. You can do this with the **Smart Tool** property or by using the **Select Tool** property to highlight an area around the edit point and pressing *F* on your keyboard.

14. Repeat *steps 9 to 13* until you've removed all the errors.

How it works...

This method relies on the *down* arrow moving the insertion point to the playhead while it's still playing. You could stop the playback, rewind it with the nudge tools, and precisely add an edit point where you want, but I find this constant stop-and-start motion mostly slows the flow of editing down. I prefer to listen to the entire track, add the edit points, then remove the clips that contain errors after the fact.

Navigating back and forth in the timeline with the *L* and *'* keys helps you quickly move to different edit points while holding *Shift* selects those offending clips. With the session in Shuffle mode, the clips will automatically "snap" to each other as you delete portions of the clips.

Keep in mind that this method only works well with a single track of audio since removing sections from multiple tracks could throw things out of sync or potentially remove important dialogue from another track. With multiple tracks, creating a group that edits all the tracks together and then toggling it on/off as you make changes can assist with solving that issue.

The most time-consuming part of this method is going back and removing the bad clips, along with fine-tuning the edits. The next recipe will show you a different way to edit that removes most of this step.

Editing during playback

The previous recipe showed you how to quickly add edit points while listening to playback, but it requires that you go back and remove clips afterward. In this recipe, we'll look at a way to edit audio clips as the clips are playing back, and a different way to align them all back together.

Getting ready

For this recipe, you will need a Pro Tools session with an audio track placed in a track. If you want to try this recipe out on a practical example, then having a recorded voice with errors can help. One has been provided in the GitHub project for this book. Otherwise, an empty audio clip or any audio clip altogether can help.

How to do it...

We're going to play back some audio and then select a portion of it that contains an error while it's playing and delete it. Then, we'll line up the clips quickly. Follow along with these steps:

1. Activate the **Select Tool** property and place the insertion point at the beginning of the clip you wish to edit.

2. Begin the playback by pressing the spacebar on your keyboard or the transport.

3. As the audio plays back, listen for errors – when you identify one, use the **Select Tool** property and click and drag over the mistake, as well as any silence or errors that fall after it.

4. Press *Backspace/Delete* to remove the clip.

5. Continue listening to the clip, highlighting issues and removing them as it plays back.

6. When you're done, stop the playback by pressing the spacebar on your keyboard or the transport.

7. Press *Command + A* (*Ctrl + A* on Windows) to select all the clips in the track.

8. In the **Clip List** area at the top right, click the arrow to bring up the clip list menu.

9. From the Clip List menu, select **Timeline Drop Order | Left to Right**.

10. Click on the arrow to bring up the Clip List menu again and select **Spot To Edit Insertion**.

11. All the clips should now be lined up, but the original clips will still be on the track – either select them and delete them or cut these clips with *X*, use the *;* or *P* key to move a track up or down, and then press *V* to paste them.

How it works...

It is possible to highlight, edit, and make changes to tracks while the audio is playing. If you make changes to timing (such as editing in Shuffle mode), then this can cause issues, so it's best to keep it in Slip mode as you perform work like this. Since Slip mode leaves the clips in their original position on the tracks, you need to line them up after the fact. You could put the session into Shuffle mode and drag the clips to the start of the track, which would snap them into place. You could also use a combination of the cursor movement keys and cutting and pasting to achieve the same result. However, these methods can become very problematic when you start working with a large number of clips.

The Clip List's **Spot to Edit Insertion** is an easy way to drop one or more clips into a session at a desired location. Depending on the Timeline Drop Order, you can have them set to **Top to Bottom**, which is good for things such as multitrack recording, but in our scenario, **Left to Right** places them all lined up in order. However, for this to work correctly, you need to edit the track in chronological order. Pro Tools appends a number to each clip, according to the number in which the edit was made, so if you go back and add an edit that's earlier in the chain, it might be placed out of order later. You can also account for this by changing the sorting method. In the **Clip** menu, select **Sort by | Start in Parent** to place the clips in the order in which they existed in the original clip.

Editing while recording

A major shoutout goes to Oleg Pertsovsky and Slava Nesmeyanov, who developed and shared this technique with me (and permitted me to share it with you in this book). Editing while recording a track is typically not possible. In the previous recipe, we showed you how to edit a track as you listen to it, but due to the way audio is captured to disk, it is not possible to do the same thing to a track while it is being recorded. There are ways of achieving the same result using an alternate track as a proxy for your edit points. We'll go over how to do this in this recipe.

Getting ready

For this recipe, you will need a Pro Tools session with three mono audio tracks. We will be performing the work as if you are recording someone as they speak, so you may do this with your own microphone if you have one, or you can simply record a blank track to simulate the workflow.

The commands in this recipe require **Keyboard Focus** to be active in the **Edit** window. You can do this by clicking the small **az** icon at the top right of the **Edit** window or pressing *Command + Option + 1* (*Ctrl + Alt + Numpad 1* on Windows).

How to do it...

We're going to use a blank audio clip as a proxy for where in the recording audio should be deleted. Once we've done this, we'll use the Grabber Tool's object mode to transfer those edit points to the recorded track. Follow along with these steps:

1. On the second track in your session, highlight a selection equal to the amount you want to record with the **Select Tool** property.

2. In the menu bar, go to **Edit | Consolidate** to generate an empty clip.

3. With the empty clip still highlighted, press *C* to copy the clip, ; to move the cursor down, and then *V* to paste it into the third track:

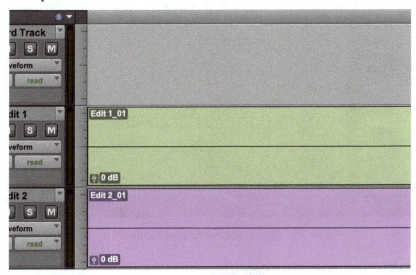

Figure 9.4 – Empty clips

4. Record arm the first track by pressing the red circle under the track's name.

5. Record arm the session by pressing the red circle in the transport and pressing play (or the spacebar on your keyboard) to begin recording.

6. As the audio is being recorded, when an error or mistake is made, highlight a selection of the second track's empty clip that corresponds with that mistake using the **Select Tool** property and press *Backspace/Delete* to remove it.

7. Continue selecting and deleting parts of the empty track that correspond to the mistakes on the recording until the recording is complete – stop the playback when you're done.

8. Make sure **Layered Editing** is disabled. This can be done in the **Options** menu or with the button furthest to the right under the tool selector in the toolbar.

9. In the toolbar, click and hold the **Grabber Tool** property and choose **Object Tool** (you can also cycle through the Grabber tools by pressing *F8* repeatedly).

10. Click on one of the clips in the second track you've been performing edits on.

11. Press *Command + A* (*Ctrl + A* on Windows) to select all the clips.

12. Click and drag these clips objects to the track below the third track, which contains the empty clip:

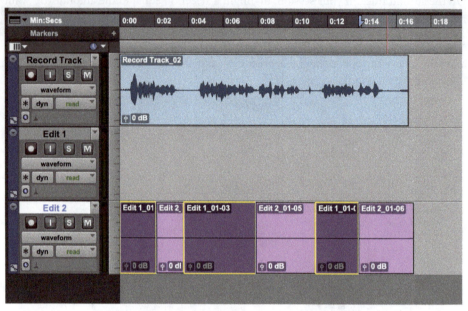

Figure 9.5 – Object clips overlapping another clip

13. Press *Backspace/Delete* to remove these clips.

14. Select one of the newly cut clips in the third track and press *Command + A* (*Ctrl + A* on Windows) to select all of these clips.

15. Drag the clips to the record track.

16. Change the edit mode to **Shuffle** by clicking on the **Shuffle** button at the top left of the **Edit** window or by pressing *F1* – you will see a warning about not being able to select the **Object Grabber** tool in **Shuffle** mode. Ignore this warning and click **OK.**

17. Press *Backspace/Delete* to remove all the mistakes and have your clips lined up automatically.

How it works...

This technique uses Pro Tools' **Object Grabber** to work around not being able to delete audio as it's being recorded. **Object** mode selects the clips as separate regions in the track, as opposed to using the standard **Grabber Tool** property or even the **Selector Tool** property and highlighting between clips. The standard behavior in Pro Tools is to select all the data between clips, including areas where no audio clip is present. If you were to select all with the standard **Grabber Tool** and drag that selection, this would result in empty spaces between the clips also being moved. In this case, the entire workflow wouldn't work.

Deleting selections in the empty clip leaves you with regions representing what you want to keep, so we need to invert the selection. This is why you need another track with an empty clip generated. When **Layered Editing** is disabled, moving one clip over another overwrites this clip data. Deleting the selected clips leaves you with regions you can move to the record track to delete. If you have **Shuffle** mode enabled when you delete these objects, all the items will line up, and the most difficult aspects of editing will be done. Some fine-tuning will still probably be needed.

This method can also be applied to pre-recorded tracks if you're editing someone else's recordings. Compared to the previous methods, I find this method much faster and less tedious.

The Punch and Roll method to record voice

While it's important to be able to edit voice work after it's recorded, it's always useful to have recordings already edited before you try to incorporate them into a project. Before the advent of digital recording tools, analog recordings and mixes benefited from Punch and Roll methods, also known as Rock and Roll (as in rock the track back a bit, then roll the recording). This is how worked: when recording a track and a mistake was made, the recording was stopped, rewound a few seconds, and then the recording was played back for the talent to find their "groove" and rhythm with what they were performing. When the point before the mistake was made is reached, the track is switched to record arm mode and the talent continued their performance. This was far more effective than recording long takes on relatively expensive analog tape and trying to edit it by literally cutting and taping afterward. What you were left with was a track ready to be mixed.

We're going to emulate this method in Pro Tools and help you record yourself or your talent, which will leave you with an audio track that's already been edited.

Getting ready

For this recipe, you'll need a blank Pro Tools session with a single mono audio track. Since the core of this workflow involves recording voice, it is very beneficial to have a microphone connected to Pro Tools to capture your voice.

For this method to work, you will need to make sure that **QuickPunch** and **Low Latency Monitoring** are enabled in the **Options** menu.

Finally, we'll be using the nudge command to rewind the audio when a mistake is made. How far back you rewind is a matter of personal preference, but I prefer 5 seconds. You can change the value in the nudge window next to the transport in the toolbar.

How to do it...

We're going to record some voice work, stop when a mistake is made, then roll the track back and punch in to overwrite the error. Follow along with these steps:

1. Set your record track to the record arm by pressing the red circle under the track name.
2. Press *Numpad 3* to begin recording.
3. When a mistake is made, press the spacebar to stop recording.
4. Press , to jump the nudge value back.
5. Press the spacebar to resume the playback. Once you have located the spot just before you made the error, press *Numpad 3* to resume recording.
6. Continue recording and punching in as needed until your track is complete; you may want to go back and clean up any edits.

How it works...

Just like the analog method described at the beginning of this chapter, using the Punch and Roll method is a matter of stopping the recording, rolling back so that you have time to preview it, and then setting the record arm when you hit the point where you want to record again. With spoken word projects, it's beneficial for the talent to begin speaking before recording to get back into the rhythm of the performance. By hearing their original recording and themselves, it's easier to match the intonation and pace of what they are saying.

Locating the correct spot to punch in becomes a bit of an art as well. Usually, you will aim for brief pauses in a sentence, or where the inflection of sentence changes, such as where a comma is written. Keeping **Layering Editing** enabled will also retain previous takes so that you can fine-tune the edit points afterward if needed.

For this to work, **Low Latency Editing** should be enabled. What this does is mute the audio passthrough that's being recorded. So, you can still listen to the audio that was previously recorded, but once you hit record as it's playing, it won't introduce that roundtrip playback and cause a delay that will cause issues with the talent.

If you are working with a keyboard without a number pad, the same methods can be performed with the transport in the toolbar. Begin the playback with the transport or the spacebar, then click the **Record** button in the transport at the punch-in point.

Using (reverse) noise gates to fill dead air

Special thanks to Christopher Johnson for demonstrating this technique and allowing me to share it in this book.

While noise reduction tools have become super effective in recent years, it's typically a good idea to leave a bit of ambient noise on a voice track. Too much noise reduction can sound harsh and unnatural, and audiences expect to hear some kind of "air" sound on a track. You could edit in noise to mask silence, but that can be time-consuming. You could also have a separate noise track running constantly, but that might mean having to bring up the overall noise print. Instead, this method uses a side-chained compressor to lower a noise track while there is a voice playing, then increases it when there is silence.

Getting ready

For this recipe, you will need a Pro Tools session with two mono audio tracks, one with a voice clip and another with a noise print. You can use room tone from a library, but ideally, the noise should be sourced from the voice track. If you have a dedicated noise print recorded in the same space, you can use this (if you get to work during a recording session, it's always helpful to dedicate some time to recording room tone). Otherwise, you can copy and paste a section of silence captured on the voice track. Either way, you can use the **Loop** tools to extend the clip to the length of the track you need.

In your Pro Tools session, make sure that the **Sends** and **Inserts** columns are visible in the **Edit** window. You can do this by going to the menu bar and selecting **View | Edit Window Views** and making sure the appropriate items are checked, or by using the **Edit Window View selector** dropdown directly above the track headers on the left-hand side of the **Edit** window.

How to do it...

We'll send the signal from a voice track to a compressor via a sidechain to lower its volume while it's playing back. Follow along with these steps:

1. On your voice track, click the first Send slot and select **bus | Bus 1**.
2. Hold *Option/Alt* and click on the send fader that appears to set it to 0 dB.
3. On the noise track, click the first insert and select **plug-in | Dynamics | Dyn3 Compressor Limiter**.
4. At the top left of the plugin window, click the drop-down menu next to the key symbol that reads **no key input** and select **bus | Bus 1**.

5. Set the plugin's parameters to the following values:

    - **RATIO**: `20.0:1`

    - **THRESH**: `-30.0 dB`

    - **ATTACK**: `1.0 ms`

    - **RELEASE**: `100 ms`

    - **GAIN**: `0 dB`

6. Start the playback with the transport or press the spacebar and listen to the results. Adjust the values of the plugin and the volume of the noise track until you get a result you enjoy – the knobs with the most impact are the attack and release to adjust how quickly the noise is introduced and lowered.

How it works...

This recipe's concept is similar to the mixing methods described earlier in this chapter. In this case, instead of having the music and other tracks react to the voice track, the noise track increases and decreases appropriately. This workflow is best suited for specific types of projects, typically long-form projects with single voices, such as audiobooks.

Appendix

Audio library management tools

Like any software, working with Pro Tools requires discovering and working within the constraints of what the software manufacturers envision and implement. When it comes to audio library management and traversal, this is one of the areas where external tools can be more functional.

Here are some popular sound library management tools available at the time of this writing.

Soundminer

One of the earliest examples of audio file search software, since 2002, Soundminer has made significant improvements and added many innovations and features to help sound editors and designers. These tools interface with Pro Tools and other DAWs allowing you to not only search for audio according to filename and metadata but also drop files directly into a session. Soundminer also includes Radium, a sampler designed to help create unique sounds from within your library. There is also a server version of Soundminer that allows a library to be stored in a central location to be accessed by multiple users. While a super powerful tool, Soundminer also has a high cost associated with it ($899 at the time of this writing). There are more affordable versions available as well with features removed. Soundminer is found at `https://store.soundminer.com/`.

BaseHead

More affordable than Soundminer, BaseHead boasts many of the features of its competitor. BaseHead has robust features and the ability to preview and spot clips directly to projects too, so it's a matter of looking at the features and their interfaces to decide which one would work best for you. Like most of the software on this list, you can try BaseHead for free to see whether it will work for you and your workflow. Basehead is available at `https://baseheadinc.com/`.

Soundly

A subscription-based service, Soundly is designed mostly to be used for its own library of sound (consisting mostly of sound effects) but it can be used for local libraries as well. There is still metadata support and Pro Tools integration, so this one may be worth checking out. There is also a free version that can import up to 10,000 local sounds. Soundly can be found at `https://getsoundly.com/`.

SoundQ

From Pro Sound Effects comes another subscription-based software for accessing their sound effects library, but it comes with many features similar to the other apps mentioned here, such as metadata and spotting directly to Pro Tools. This one is good for audio engineers and sound designers looking to tap into Pro Sound Effect's existing library. SoundQ is available at `https://www.prosoundeffects.com/soundq/`.

Sononym

A unique offering at a somewhat more affordable price (one-time purchase of $99 at the time of this writing), Sononym offers a different way of searching sounds with its machine learning approach. Rather than relying exclusively on metadata and filenames, Sononym tries to find similar-sounding files and present them to you. This also allows for some unique filtering options to try and help narrow down the exact sound you are looking for. You can find Sononym at `https://www.sononym.net/`.

There's more

This is by no means an exhaustive list, but it can steer you in the right direction for audio library management. One thing that is important to know about regardless of which software you might use is the **Universal Category System (UCS)**.

UCS

Created by a team of audio engineers, UCS provides a public domain document that guides audio engineers and editors with a unified way of categorizing sound effects. Prior to UCS, each sound library publisher and audio professional would have their own categories, and trying to locate and organize sounds could be very frustrating. Most of the libraries previously listed include UCS integration, making it easier and more consistent to find sounds across different libraries, but you can also access and learn all the different conventions on their site: `https://universalcategorysystem.com/`.

Kraken

Kraken is a dialogue-specific tool that is geared toward dialogue editors to help organize and find alternate takes and expedite assembly workflows. It's worth checking out if motion picture audio and dialogue editing is your area of focus. Kraken is available at `https://krakensoftware.co.uk/`.

What's next?

This book has demonstrated workflows and techniques for Pro Tools that I've developed over the years, but the real test comes in putting them into practice. When working on your next project, consider the time you are spending on a specific task. Think about what steps you are taking to complete that task, and whether there are any opportunities to improve it. Review the recipe titles in this book and see whether there are ways you can apply the techniques outlined to your own work.

Perhaps you will find the methods I've outlined improve your workflow, or perhaps you find something that doesn't quite fit or causes a stumbling block. This is your opportunity to put your own personal touches into that recipe. Much like following a recipe from a cookbook, they often serve as guidelines that you can modify and adjust to your liking. Perhaps you prefer interfacing with Pro Tools in a more visual way; then use the on-screen tools as opposed to the keyboard shortcuts. Maybe you've already found a method in your workflows that you've grown accustomed to over time; then these recipes can be used to augment them.

The most important thing is to be open to learning new things. I've been using Pro Tools for 25 years and I learn something new about it every day. Even if you don't upgrade the software, I truly think it's impossible to know everything about it – I don't even think there is a person at Avid who does. The more you use Pro Tools, the better you will get at it. As your skill and proficiency increase, you'll be able to spend less time with the technical aspect of the software and instead focus on the creative output you can achieve with it.

Index

Packtpub.com

Subscribe to our online digital library for full access to over 7,000 books and videos, as well as industry leading tools to help you plan your personal development and advance your career. For more information, please visit our website.

Why subscribe?

- Spend less time learning and more time coding with practical eBooks and Videos from over 4,000 industry professionals

- Improve your learning with Skill Plans built especially for you

- Get a free eBook or video every month

- Fully searchable for easy access to vital information

- Copy and paste, print, and bookmark content

Did you know that Packt offers eBook versions of every book published, with PDF and ePub files available? You can upgrade to the eBook version at packtpub.com and as a print book customer, you are entitled to a discount on the eBook copy. Get in touch with us at customercare@packtpub.com for more details.

At www.packtpub.com, you can also read a collection of free technical articles, sign up for a range of free newsletters, and receive exclusive discounts and offers on Packt books and eBooks.

Other Books You May Enjoy

If you enjoyed this book, you may be interested in these other books by Packt:

The Music Producer's Creative Guide to Ableton Live 11

Anna Lakatos

ISBN: 9781801817639

- Understand the concept of Live, the workflow of recording and editing Audio and MIDI, and Warping.

- Utilize Groove, MIDI effects, and Live 11 s new workflow enhancements to create innovative music.

- Use Audio to MIDI conversion tools to translate and generate ideas quickly.

- Dive into Live's automation and modulation capabilities and explore project organization techniques to speed up your workflow.

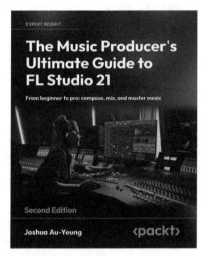

The Music Producer's Ultimate Guide to FL Studio 21

Alexander Zacharias

ISBN: 9781837631650

- Get up and running with FL Studio 21.

- Compose melodies and chord progressions on the piano roll.

- Mix your music effectively with mixing techniques and plugins, such as compressors and equalizers.

- Record into FL Studio, pitch-correct and retime samples, and follow advice for applying effects to vocals.

- Create vocal harmonies and learn how to use vocoders to modulate your vocals with an instrument.

Packt is searching for authors like you

If you're interested in becoming an author for Packt, please visit `authors.packtpub.com` and apply today. We have worked with thousands of developers and tech professionals, just like you, to help them share their insight with the global tech community. You can make a general application, apply for a specific hot topic that we are recruiting an author for, or submit your own idea.

Hi!

I'm Emiliano Paternostro, author of *The Pro Tools 2023 Post-Audio Cookbook*. I really hope you enjoyed reading this book and found it useful for increasing your productivity and efficiency in Pro Tools.

It would really help us (and other potential readers!) if you could leave a review on Amazon sharing your thoughts on it.

Please visit the link below or scan the QR code to leave your review:

`https://packt.link/r/1803248432`

Your review will help us to understand what's worked well in this book, and what could be improved upon for future editions, so it really is appreciated.

Best wishes,

Emi

Download a free PDF copy of this book

Thanks for purchasing this book!

Do you like to read on the go but are unable to carry your print books everywhere? Is your eBook purchase not compatible with the device of your choice?

Don't worry, now with every Packt book you get a DRM-free PDF version of that book at no cost.

Read anywhere, any place, on any device. Search, copy, and paste code from your favorite technical books directly into your application.

The perks don't stop there, you can get exclusive access to discounts, newsletters, and great free content in your inbox daily

Follow these simple steps to get the benefits:

1. Scan the QR code or visit the link below

https://packt.link/free-ebook/978-1-80324-843-1

2. Submit your proof of purchase
3. That's it! We'll send your free PDF and other benefits to your email directly